Teaching Mathematics in Colleges and Universities: Case Studies for Today's Classroom

Faculty Edition

Selected Titles in This Series

CBMS

Conference Board of the Mathematical Sciences

Issues in Mathematics Education

Volume 10

Teaching Mathematics in Colleges and Universities: Case Studies for Today's Classroom

Faculty Edition

Solomon Friedberg

Avner Ash
Elizabeth Brown
Deborah Hughes Hallett
Reva Kasman
Margaret Kenney
Lisa A. Mantini
William McCallum
Jeremy Teitelbaum
Lee Zia

American Mathematical Society
Providence, Rhode Island
in cooperation with
Mathematical Association of America
Washington, D. C.

2000 *Mathematics Subject Classification.* Primary 00A35, 97D40; Secondary 00A05, 97C70, 97D30, 97D50, 97D60, 97D70, 97U70.

Contents

Part II. Supporting Materials for Faculty

Acknowledgements

It is a pleasure to thank the many people who have enabled this project to take place. When I first had the idea of writing case studies for mathematics graduate students and instructors, I discussed it with Deborah Hughes Hallett and Katherine Merseth. Their thoughtful and enthusiastic responses were key to making the idea a reality. Deb and Kay also contributed their expertise in helping me formulate a realistic plan of action. At the grant-writing stage, Avner Ash, Margaret Kenney, Jim Leitzel, Glenn Stevens, Lee Zia, and Dorothy Wallace, as well as Deb and Kay, stepped in with valuable comments which improved the development plan and the grant proposal considerably. Thank you all. Jim Leitzel's untimely death saddened all of us greatly; his memory inspires us still. A host of Boston College administrators gave me the green light early on to pursue this idea, and they ultimately committed extensive resources to support the project: Michael Smyer, Vice President for Research, Fr. Robert Barth S.J., former Dean of the College of Arts and Sciences, Joseph Quinn, present Dean, and Richard Jenson, Chair of the Mathematics Department. In addition, John Neuhauser, both as Dean of the School of Management and later as Academic Vice-President and Dean of the Faculty at Boston College, shared his extensive experience with the use of case studies, and made a difference. No faculty member could ask for better support from his institution. My proposal to carry out this project was funded by the Fund for the Improvement of Postsecondary Education (FIPSE), a unit within the Office of Postsecondary Education, U.S. Department of Education, under FIPSE grant number P116B980015, and I hereby acknowledge with appreciation this support.

In carrying out the project, The Boston College Mathematics Case Studies Project (BCCase), development team members Avner Ash, Elizabeth Brown, Deborah Hughes Hallett, Reva Kasman, Margaret Kenney, Lisa Mantini, William Mc-Callum, Jeremy Teitelbaum, and Lee Zia contributed their ideas, energy, experiences, and wisdom, and made the project what it is. A enormous, sincere, and much-deserved thanks. Elin Norberg and David Foster ably served as the project administrators, and Mary Sullivan contributed her expertise as the project evaluator. Their contributions are much-appreciated.

I and the rest of the BCCase team would like to thank the many people who have aided us in the development of the case studies. Once the project got under way, we depended upon the feedback and insightful comments of mathematics graduate students from around the country who used and evaluated the case studies as we wrote and revised them. We would like to sincerely thank them all. We would like to especially mention the group of graduate students whose initial use often led to the early demise or major revision of a case. The group included: Elizabeth Brown, Ben Brubaker, Mark Evans, Craig Friedland, Lynette Kelley, Sarah Lehan, Brian Munson, Allison Pacelli, Carolyn Pointek, Steven Rattendi, and Jay Douglas Wright. Other contributors include: Dawn Berk, Susan Billimek, Jonathan Cox, Teodora Cox, Jailing Dai, Jeff Edmunds, Todd Grundmeier, Tom Harris, Michael Hayes, Jessica Hemenway, Jack Hoppin, David Hrencecin, Sarah James, Katrina Jimenez, Qayum Khan, Michael Kuecken, Amy Lehan, Andre Lehovich, Arthur Lo, Guada Lozano, Joyce Macabéa, David Marsden, Kaarin McCarthy, Laura McSweeney, Stephanie Molnar, Carlos Morales, Kinya Ono, Andy Parker, Virgil Pierce, T.J. Preacher, Chris Rasmussen, Randi Scott, Charlotte Schulze-Hewett, Jeff Selden, Randy Sesto, Jeanine Smallwood, Jennifer Smith, Nichole Soter, Carrie Spooner, Simei Tong, Nadia Whisenand, Haishen Yao, and Shaowei Zhang. Also thanks to the many graduate students from around the country who filled out copious feedback forms after using a case and offered their frank comments and intelligent suggestions. Parallel to the contributions of the mathematics graduate student community has been the contribution from mathematics faculty. We would like to thank the small army of faculty from around the country who have tested and evaluated the case studies or provided other feedback to us: Doug Aichele of Oklahoma State University, Judith Arms of the University of Washington, Margaret Balachowski of Michigan Technology University, Ken Boggart of Dartmouth College, Tina Garn of the University of Arizona, Thomas Goodwillie of Brown University, Daniel L. Goroff of Harvard University, Marcia Groszek of Dartmouth College, Tim Gutman of the University of New Hampshire, Gary Harris of Texas Tech University, Diane Herrmann of the University of Chicago, Theodore Laetsch of the University of Arizona, Katherine Merseth of Harvard University, Teri Jo Murphy of the University of Oklahoma, Regina Panasuk of the University of Massachusetts Lowell, Kent Pearce of Texas Tech University, Emma Previato of Boston University, James Propp of the University of Wisconsin, Karen Rhea of the University of Michigan, David Rohrlich of Boston University, Ned Rosen of Boston College, Eileen T. Shugart of Virginia Tech., Brenda Slez of the University of Massachusetts Lowell, Glenn Stevens of Boston University, Maria Terrell of Cornell University, Stuart Thomas of the University of Oregon, Dorothy Wallace of Dartmouth College, Steve Wheaton of the University of Arizona, and Dale Winter of Harvard University.

We would also like to thank Marilyn Adams, the Boston College mathematics department secretary, Susan Hoban and the rest of the staff of the BC Office of Research Administration, Jay Donahue, our program officer at FIPSE, Naomi Fisher and Bonnie Saunders of the MER Forum, Elizabeth Armstrong and Tom McGarry of the Harvard Medical School, Tom Seidenberg and Gwen Sneedon of the Phillips Exeter Academy, and Edward Dunne and Barbara Beeton of the American Mathematical Society.

Finally, I would like to thank my wife Karen Ann Siller and my children Ilana, Rina, and Liora, who saw me come home late or work on the computer so many evenings due to this project, who put up with my trips to conferences and case study presentations as I worked to introduce these materials to the mathematics community, and who were always there to greet me with open arms and big smiles upon my return.

Thanks to all.

Solomon Friedberg
Director, The Boston College Mathematics Case Studies Project
Chestnut Hill, Massachusetts
February 12, 2001

Introduction

Progress in mathematics frequently occurs by first studying particular examples and then generalizing the patterns which have been observed into far-reaching Theorems. Similarly, in teaching mathematics one frequently employs examples to motivate a general principle or to illustrate its use. This volume employs the same idea in the context of learning *how* to teach: by analyzing particular teaching situations one may develop broadly applicable teaching skills useful for the professional mathematician. These teaching situations are the Case Studies of the title. Just as a good mathematician seeks to understand the details of a particular problem but also to put it in a broader context, the examples presented are chosen to offer a serious set of detailed teaching issues but also to afford analysis from a broad perspective.

Why use examples to develop teaching skills, rather than simply giving general principles? One reason is that it is difficult to learn teaching solely from such principles. Just as 'doing the exercises' is an integral part of learning mathematics (if the exercises are well-conceived rather than busy-work), these Case Studies may be regarded as teaching exercises, and can play a similar role in gaining teaching expertise. A second is that no two people have the exact same idea of what good teaching actually is— in contrast to mathematics, there is frequently no one right answer. Even highly regarded teachers possess different skills and achieve different outcomes; one may enable the better students to perform at a very high level, while another shows the weaker students that, for the first time in their lives, they can do mathematics. Similarly, there is no one right answer to the Case Studies presented here. In other words, principles of good teaching are personal, and the goal here is for each person to critically develop such principles, but not to arrive at the same set of them. Finally, in teaching every day is different. To be a successful teacher, it is important to be able to analyze and deal with classroom situations as they develop. The Case Studies prepared by this project present a broad range of teaching scenarios, and give participants the opportunity to think them through. Doing so will help prepare for the next, once again different, classroom experience.

One aspect of good teaching is technical: write legibly, use the board effectively, speak audibly. These Cases do not address these issues. Rather, their focus is on more conceptual issues, in the broad areas of mathematical content as perceived by the students, of pedagogy, and of faculty-TA relations. For example, how does one help students to truly master the big ideas, such as the derivative, the integral, and the relation between them?

1

Manage a classroom of students with a wide range of background knowledge and of ability? Balance teaching and other responsibilities, such as completing one's dissertation?

Finally, just what is a Case Study, and what does one do with it? A Case is an excerpt from a teaching situation, described from the perspective of various students and of the instructor. The Case raises a variety of pedagogical and communication issues, to be explored and analyzed in group discussion, for example by a group of graduate teaching assistants together with a faculty facilitator. Group consideration of a Case offers the advantage of drawing upon collective experience and diverse perspectives, and allows different issues, ideas, and strategies to be considered and discussed. The methodology of Case Studies is widely used in this way in areas such as business and law, and also in teacher development, both university-level and precollegiate, in diverse subjects from the humanities to accounting. Alternatively, a reader working independently may take each Case as an exercise, thinking about the situation, asking what the different issues are, what he or she would do next or would have done differently, what can be learned.

We hope that the consideration of these mathematics Cases, in either a group or an individual setting, will be thought-provoking, and will help each reader to develop high-quality teaching skills for use in his or her own classroom.

Part I

Fourteen Case Studies

Changing Sections

Date: Mon, 23 Aug 1999 09:11:04 -0700 <MST>
From: Otto Vorsky <ottov@math.state.edu>
To: wgm@math.state.edu
Subject: My Math 125 class

Dear Professor Maddox,

I just finished giving my students in Math 125 a quiz on precalculus material. Some of them are really weak; they couldn't even solve $\ln(x + 5) = 10$ (one of them thought that $\ln(x + 5) = \ln(x) + \ln(5)$!). I told them they should drop the class and take precalculus instead. O.K.?

Otto

Date: Tue, 24 Aug 1999 16:50:23 -0700 <MST>
From: Felicia Lopez <flopez@math.state.edu>
To: Walter Maddox <wgm@math.state.edu>
Subject: student wanting transfer into my Math 125 section

Dear Professor Maddox,

I just had a long conversation with a student wanting to transfer into my section of Math 125. He was very upset; apparently he was advised by his TA to drop 125 and take precalculus instead. He says that he did very well in precalculus in high school, but wasn't prepared for the pop quiz the TA gave them on the first day of classes, and made a lot of mistakes. He thinks he could catch up. I tested him on a few things. He isn't too bad on basic algebraic manipulations, but has a lot of trouble with logs and trig.

I'd be willing to take him in my section and work with him to help him catch up, but I'm at the upper limit of 35 (I've had a whole rash of transfers over the last couple of days for some strange reason). I talked to one of the secretaries, and she says I need your permission to raise the maximum enrollment. The student is returning on Friday to find out what he should do.

Yours sincerely,

Felicia Lopez

Date: Wed, 25 Aug 1999 10:10:03 -0700 <MST>
From: Felicia Lopez <flopez@math.state.edu>
To: Otto Vorsky <ottov@math.state.edu>
Subject: thanks (not)

Hey Otto,

I've had four students switch from your section to mine. What are you doing to them? I can't take any more; I'm at my maximum enrollment already and had to ask Professor Maddox for permission to add the last one. (I didn't let on that all the other ones came from your section . . . thank me.)

Felicia

Date: Wed, 25 Aug 1999 11:37:28 -0700 <MST>
From: Otto Vorsky <ottov@math.state.edu>
To: flopez@math.state.edu
Subject: Re: thanks (not)

Dear Felicia,

I gave my students a pre-calculus test the first day of classes, as I usually do. Some of them did quite poorly. One thought that $\ln(x + 5) = \ln(x) + \ln(5)$. When I told him that the rule was $\ln(xy) = \ln(x) + \ln(y)$, he seemed to think the difference was trivial, and I shouldn't penalize him for such a small error. I really felt obliged to let him know now that with his weak background he wouldn't be able to survive in my class.

I'm sorry that they are all running to you; it's because you're known as such a good teacher. You really don't have to accept them; I think it would be best for them in the long run if they took the prerequisite courses again and learned the material properly.

Regards,

Otto

Date: Wed, 25 Aug 1999 13:42:59 -0700 <MST>
From: Felicia Lopez <flopez@math.state.edu>
To: Otto Vorsky <ottov@math.state.edu>
Subject: Re: thanks (not)

I think the student you're talking about (who had trouble with the logs) is the one who came to see me yesterday afternoon. You know, he was really upset. I agree that he ought to be better prepared, but now that he's here shouldn't we try to help him? Besides, he's not as bad as you make out; we went through your quiz together and he recognized a lot of his mistakes.

I think he's like a lot of our students; he's capable of doing the algebra, but it doesn't seem to stick in his head for very long. I don't see the point in sending him back to a course that has already failed him once. I usually do a bit of precalculus review throughout the semester; I'll just give him some extra worksheets.

Felicia

Date: Thur, 26 Aug 1999 07:36:45 -0700 <MST>
From: Otto Vorsky <ottov@math.state.edu>
To: flopez@math.state.edu
Subject: student

Dear Felicia,

Maybe the course failed him, maybe he had a poor teacher, maybe he didn't work hard enough, or maybe he just needs to see the material twice for it to sink in. I don't know. All I know is that he can't proceed any further without knowing the prerequisites.

I don't do any remedial work in my section—I make the prerequisites clear and emphasize that if they don't know something from precalculus then it is their responsibility to learn it. This gives me more time to talk about calculus, which is what the rest of the class is there to learn. Last year my section had the highest average on the final.

ERROR - UNABLE TO FIND SERVER - THE PREVIOUS MESSAGE WAS NOT DELIVERED

Date: Thur, 26 Aug 1999 08:07:00 -0700 <MST>
From: Walter Maddox <wgm@math.state.edu>
To: Felicia Lopez <flopez@math.state.edu>
Subject: enrollments

Dear Felicia,

I think 36 is too large for your section (35 is too large, but that seems to be a fait accompli). So I think you should tell this student that he cannot switch to your section.

You are being diplomatic but this student must be switching from Otto's section. Otto seems to have this effect on students in the first week or so. I'll write to him and tell him to ease off.

Walter

Date: Thur, 26 Aug 1999 12:27:06 -0700 <MST>
From: Otto Vorsky <ottov@math.state.edu>
To: Felicia Lopez <flopez@math.state.edu>
Subject: Your student

Dear Felicia,

I tried to reply earlier but was having trouble with my email. You shouldn't necessarily blame the student's previous math courses; he has to take some of the responsibility himself. There's a tendency in the U.S. to go too far in accommodating unprepared students. I don't think this is doing them any favors; if we don't enforce the prerequisites now it will just hurt them later.

I agree that this particular student is not as bad as I had originally thought, however . . . it turns out that he's the guy from the computing center who came round to fix the computer in our office!

I had tried to reconstruct my internet configuration after a hard disk crash last night. Apparently I mixed up the names of the outgoing and incoming mail servers, whatever they are. (It seems like a pretty silly system). I was pretty impressed with his computer expertise . . . but his mathematical abilities are lacking, and I still think that he should go back and learn the material right.

Regards,

Otto

Date: Thur, 26 Aug 1999 14:12:06 -0700 <MST>
From: Walter Maddox <maddox@math.state.edu>
To: Otto Vorsky <ottov@math.state.edu>
Subject: Re: My Math 125 class

Dear Otto,

Sorry to be so long in replying to your message—I've been swamped. It's OK with me if you want to give the students a quiz on precalc material but you shouldn't scare them into dropping your course. They switch to Felicia's section and it isn't fair to her. Just do the best you can with the students you've got.

Walter

Date: Fri, 27 Aug 1999 10:00:21 -0700 <MST>
From: Gil Roberts <groberts@u.state.edu>
To: Felicia Lopez <flopez@math.state.edu>
Subject: Our meeting today

Dear Ms. Lopez,

I'm sorry there's not room in your section. I guess I'll have to consider whether I want to deal with Mr. Vorsky, or just drop Calc and take it next semester.

Gil

Emily's Test

Emily Seitz was a nervous wreck. What she had just seen called for immediate action and she didn't know what to do. During the training seminars for graduate teaching assistants she had been forewarned about the proper behavior to follow while proctoring a test. She had been told to stay alert and keep her attention on the class while the test was in progress. The advice given was clear: "keep your eyes riveted on your students and definitely do not use the hour to get involved in doing some of your own work". In fact, it was a good idea to keep moving about the room during the entire period and to refrain from sitting down.

Although she had made no objections in response during their meeting, Emily felt her advisor had been overreacting and that it was truly not necessary to patrol a classroom of thirty-five students. But this was her first experience in proctoring and who was she to question the advice offered by a seasoned faculty member. However, in all her years as an undergraduate she had never been in a class in which cheating had been an issue. She had never even heard of cheating going on in other courses at her college. When she offered this remark to her advisor, Emily was told that perhaps she was being a bit naive about the behavior of college students.

Now, here she was in a confrontational situation. While walking up one of the aisles, Emily saw some loose pages protruding out from under the blue book of one student. In giving directions at the start of the test, Emily had told the students explicitly to use the back pages of the exam book for scrap work and that no other papers were to be used for this purpose. All that was needed was a copy of the exam and the blue book. As she passed by Liam O'Neill's chair, the student made no attempt to hide the pages with his arm. Instead, Liam was writing away apparently oblivious to her presence. She focused more closely on the sheets of paper and saw that the one on the very bottom was somewhat askew and that it clearly contained some small neat handwriting. This contrasted with the scribbled writing currently being entered in the blue book. The loose paper certainly did not appear to be a page of scratch work. However, it was also apparent Liam was not using the paper at present; he seemed unaware that it was even on his desk. Emily wondered what information was on this paper – and if he had been using it earlier on when she was in another part of the room.

Emily realized that she did not know her new university's procedures for dealing with cheating. Then she remembered a story she had once heard, a story about a graduate TA who had confronted a cheater, only to himself be accused of improper behavior. Though she was utterly mystified as to how this could have occurred, the thought tied her stomach in knots.

11

As she agonized over what to do, Emily looked up and saw a student in the next aisle watching her. Her eyes seemed to be traveling back and forth between Emily's eyes and the piece of paper under the exam book. The student, Tabatha, always made her uncomfortable. Tabatha was a complainer and whether it was about a homework grade or something going on in class that she didn't understand, she had an attitude. At this moment, Tabatha had a knowing smirk on her face. Emily felt as if she were the one being tested. If she did the right thing according to Tabatha, all would be well. On the other hand, if she didn't, she was certain Tabatha was prepared to act. What should Emily do?

Fundamental Problems

Part I

Keng was the TA for a large calculus class at Western University. Over 80 students attended the professor's lectures 3 times a week, while they had section once a week with Keng in groups of 25-30. Since there was a midterm in two days, Keng had agreed to run a review session for the whole class this evening. He wasn't used to teaching so many students at once, but several students had told him in office hours that they were confused by the professor's lecture on the Fundamental Theorem, and had told him how much they would appreciate a review session before the exam.

Keng began by saying "Okay, this midterm covers Chapter 5, which is about the integral. I think the three most important things we've learned in Chapter 5 are One, what the definite integral is; Two, computing integrals by antiderivatives and the Fundamental Theorem of Calculus; and Three, how to do integrals by substitution. [Keng wrote these three topics on the board.] These are the three topics I suggest you focus on as you review for the midterm. Are there any questions about these topics or problems involving them that you'd like to go over?"

Students began to ask problems. The first was an integral by substitution from the homework. Keng solved it on the board. The second was a standard problem (from the homework and just like one in the text):

> *A population of bacteria living in a Petri dish has been treated with an experimental drug. If there are 10^6 bacteria initially and the rate of growth at time t hours is observed to be $r(t) = 10^5(1 + t^2)$, find the total number of bacteria in the Petri dish after 3 hours.*

As Keng answered this he noticed that the better students were looking bored. As an aside for them, he added, "please notice that we could not solve this problem without the Fundamental Theorem of Calculus. So we see just how important this Theorem is."

Immediately, Lindsey's hand went up. She was a student who was not afraid to speak up, and Keng had noticed that her questions were often useful in figuring out what the class understood and what it did not. Keng called on her.

"I'm really confused about what you just said. In fact, I don't get the whole Fundamental Theorem thing. I thought the integral *was* the antiderivative, and now we're integrating the growth rate to get the population. I don't see why."

A voice, male, from the back of the room added "Yeah, and what does all this have to do with the derivative of

$$\int_a^x f(t)\, dt?$$

I don't understand that part at all."

The large room became silent, and Keng wondered what he should say.

Making the Grade (College Algebra Version)

You, as the teaching assistant, are working with a class of students taking College Algebra. You have been doing a review of basic algebra and applications of algebra to solving word problems. Grade the sample student work below first on a 10-point scale, as if these were problems on a quiz or an exam, and then on a 3-point scale, as if these were problems on the homework. Write your scores next to the student work in the appropriate column.

1. Solve $2(x - 10) - (12x - 4) = 20$.   10-point scale   3-point scale

Student A:

$$2(x-10) - (12x-4) = 20$$
$$2x - 20 - 12x + 4 = 26$$
$$-10x - 16 = 26$$
$$-10x = 42$$
$$\boxed{x = -4.2}$$

Student B:

$$2(x-10) - (12x-4) = 20$$
$$2x - 10 - 12x - 4 = 20$$
$$-10x - 14 = 20$$
$$-10x = 34$$
$$x = -\frac{34}{10} = \underline{\underline{-3\tfrac{2}{5}}}$$

Student C:

$$2(x-10) - (12x-4) = 20$$
$$2x - 20 - 12x + 4 = 20$$
$$-10x - 16 = 20$$
$$-10x = 36$$
$$x = -\frac{36}{10}$$
$$x = -3 + \tfrac{3}{5} = -2\tfrac{2}{5}$$
$$\boxed{x = -2\tfrac{2}{5}}$$

2. The sum of three consecutive odd integers is 81. Find the integers.

10-point scale   3-point scale

Student A:

$$\text{Let } x = 1^{st} \text{ odd integer}$$
$$x+1 = 2^{nd} \text{ odd integer}$$
$$x+2 = 3^{rd} \text{ odd integer}$$
$$x + x+1 + x+2 = 81$$
$$3x + 3 = 81$$
$$3x = 78$$
$$x = 26$$

The integers are $\boxed{26, 27, 28}$

Student B:

$$x = \text{ odd integer}$$
$$x+2 = 2^{nd}$$
$$x+4 = 3^{rd}$$
$$x + x+2 + x+4 = 81$$
$$9x = 81$$
$$x = 9 \qquad \boxed{9, 11, 13}$$

Student C:

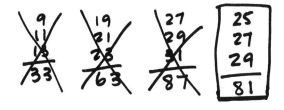

Making the Grade (Calculus I Version)

You, as the teaching assistant, are working with a class of students taking Calculus I. Grade the sample student work below first on a 10-point scale, as if these were problems on a quiz or an exam, and then on a 3-point scale, as if these were problems on the homework. Write your scores next to the student work in the appropriate column.

1. Find the derivative of $y = \sec^2(1 + 3x)$.

10-pt scale   3-pt scale

Student A:

$$y = (\sec u)^2 \quad u = 1 + 3x$$
$$y' = 2(\sec(1+3x))(\sec x \tan x) \cdot 3$$
$$= 6 \sec x \tan x \sec(1+3x)$$

Student B:

$$y' = 2 \cdot 3 \sec(1+3x) \tan(1+3x)$$
$$= 6 \sec(1+3x) \tan(1+3x)$$

Student C:

$$y = \sec(1+3x) \cdot \sec(1+3x)$$
$$\sec \tan(1+3x) \cdot 3 \cdot \sec \tan(1+3x) \cdot 3$$
$$\boxed{18 \sec \tan(1+3x)}$$

Student D:

$$y' = \tan(1+3x) \cdot 3$$

2. Let $f(x) = \sqrt{2x^2 - 4}$. Find $\displaystyle\lim_{x \to 2} \frac{f(x) - f(2)}{x - 2}$.

10-pt scale   3-pt scale

Student A:

$$\lim_{x \to 2} \frac{\sqrt{2x^3 - 4} - \sqrt{2(2)^2 - 4}}{x - 2}$$

$$\lim_{x \to 2} \frac{\sqrt{2x^2 - 4} - 2}{x - 2} = \frac{\sqrt{4}}{2} = \frac{2}{2} = 1$$

Student B:

$$f(x) = \sqrt{2x^3 - 4}$$

$$f(2) = \sqrt{2(2)^2 - 4} = \sqrt{8 - 4} = 2$$

$$\lim_{x \to 2} \frac{\sqrt{2x^2 - 4} - 2}{x - 2} \cdot \frac{\sqrt{2x^2 - 4} + 2}{\sqrt{2x^2 - 4} + 2}$$

$$= \lim_{x \to 2} \frac{2x^3 - 4 - 4}{(x-2)(\sqrt{2x^2 - 4}) + 2(x - 2)}$$

$$= \lim_{x \to 2} \frac{2x^2}{(x-2)\sqrt{2x^2 - 4} + 2x^3 - 4}$$

$$= \frac{2(2)^2}{0 + 8 - 4} = \frac{8}{4} = \circled{2}$$

Student C:

$$\lim_{x \to 2} \frac{\sqrt{2x^2 - 4} - 2}{x - 2} \quad \overset{\text{"0"}}{\underset{\text{"0"}}{}} \quad \lim_{x \to 2} \frac{(2x^2 - 4)^2 - 2}{x - 2}$$

$$\overset{L'H}{=} \lim_{x \to 2} \frac{2(2x^2 - 4) \cdot 4x}{1} = 2(8 - 4)\,8$$

$$= \boxed{64}$$

Making the Grade (Multivariable Calculus Version)

You, as the teaching assistant, are working with a class of students taking Multivariable Calculus. Grade the sample student work below first on a 10-point scale, as if these were problems on a quiz or an exam, and then on a 3-point scale, as if these were problems on the homework. Write your scores next to the student work in the appropriate column.

1. Find the tangent plane to the graph of $h(x,y) = x^2 + 3y^2$ when $(x,y) = (1,-1)$.

10-pt scale   3-pt scale

Student A:

$$h_x = 2x \qquad h_y = 6y$$
$$h_x(1,-1) = 2 \qquad h_y(1,-1) = -6$$
$$\text{tgt pl:} \quad z = 2(x-1) - 6(y+1)$$

Student B:

$$h(1,-1)=4 \qquad \frac{\partial h}{\partial x} = 2x \qquad \frac{\partial h}{\partial x}(1,-1)=2$$
$$\frac{\partial h}{\partial y} = 6y \qquad \frac{\partial h}{\partial y}(1,-1)=-6$$
$$0 = 4 + 2(x-1) - 6(y+1)$$
$$\boxed{4 = 2x - 6y}$$

Student C:

$$h(1,-1) = 4 \qquad \frac{\partial h}{\partial x} = 2x \qquad \frac{\partial h}{\partial y} = 6y$$
$$p(x,y) = 4 + 2x(x-1) + 6y(y+1)$$
$$= 4 + 2(x-1) - 6(y+1)$$

2. Let $g(x, y) = (2x - y)^2$. Sketch at least three level sets for g on the grid to the left, and sketch the graph of g.

10-pt scale   3-pt scale

Student A:

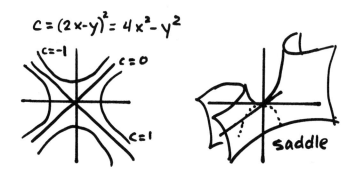

Student B:

$$c = (2x-y)^2 = 4x^2 - 4xy + y^2$$

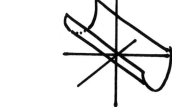

Student C:

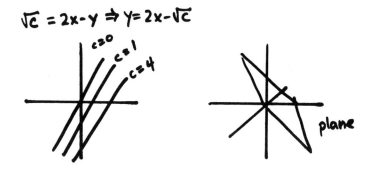

Making Waves

"You know, Kara, we're coming to the part of the Calculus course I hate most," remarked Louis as they hovered around the tea and cookies at the department Friday tea. "I've always dreaded having to cover Fourier series in this introductory class."

"There are plenty of parts of this course I find kind of tedious," replied Kara, "but I actually look forward to Fourier series. It's such a beautiful part of math, and it's something the students really haven't ever seen before – unlike, say, derivatives, which they all had in some form or other in high school."

"Sure, Fourier series are beautiful, but I don't see how they are appropriate at this level. Oh, we can explain the formulas to them, and they can calculate the first few Fourier coefficients of some simple functions, but we can't even begin to put the theory in the proper mathematical framework."

"What do you mean, proper mathematical framework?" Kara said, grinning. "It's not like we put derivatives in their proper mathematical framework. We try to illustrate what they're good for, and give some feel for how to calculate with them. Surely we can do the same with Fourier series."

"I don't know about that" said Louis. "To me, Fourier expansions are a nice example of orthogonal decomposition in Hilbert space. If you can't apply linear algebra to the picture, then I hardly see why you should bring them up at all. It's not like they have the practical applications that more basic Calculus does."

"You're kidding, right?" said Kara.

"What do you mean?"

"Fourier series have billions of practical applications. I mean, how can you understand anything about waves of any sort without Fourier series? Fourier wanted to understand heat conduction – that's why Fourier series were invented! Not to mention quantum mechanics. Surely you point this out to your students."

Louis felt somewhat affronted. He had the distinct impression that Kara was laughing at him.

"Look, I learned about Fourier series in the first year grad course on Analysis, in the context of Hilbert spaces. Fourier expansions made perfect sense to me. I believe you that the theory has applications, but I don't have the time or the training to get involved in learning them. It's different for you, maybe – you're interested in applied mathematics – but I'm working in low-dimensional topology, and quantum mechanics just isn't relevant." Even as Louis said this, he could feel that he'd come out sounding snooty, which he really hadn't intended.

"Whatever. Look, Louis, I'm not telling you to become a physicist. But I really do think you're doing your class a disservice by not giving some physical motivation for Fourier series. Plus, never mind your Calc class, you're missing a lot of interesting stuff yourself."

"OK, Kara, maybe you're right. But how exactly am I supposed to come up with this motivating material by Monday?"

"Why not stop by my class at 9 a.m.? I'm going to give my little motivating speech right at the beginning of class. Maybe that will give you some idea of what you can do at your class."

"That sounds reasonable...I don't teach until 2 p.m. OK, I'll be there Monday morning at 9."

Monday Morning

Louis slipped into the back of Kara's 9 a.m. Calculus class and tried to look inconspicuous. Sitting in the back put him among the sleepers and lost souls who congregate in the back of Calculus lectures. He was relieved when Kara ignored him and started her introduction.

"Good morning. Today we're starting a new topic, Fourier series." she said. "Fourier series are a crucial mathematical tool for the study of vibrations and waves. To give you some idea of what I mean, let's look for a moment at a guitar string stretched between two fixed points. When the guitar string vibrates, it produces a tone which we hear as a musical note. However, the guitar string can vibrate in many different ways. For example, it can vibrate up and down with just one hump:

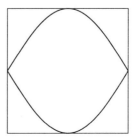

This makes a sound which we call the "fundamental tone"; it is the lowest note this length of string can produce. However, it can also vibrate with 3 or with 5 humps:

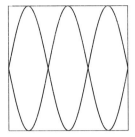 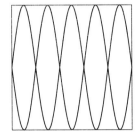

These vibrations make tones which are multiples of the fundamental tone. They sound higher than the fundamental tone; they are called "overtones" of the fundamental.

In each of these examples, the guitar string is shaped like a piece of the graph of a cosine function. In fact:

$$\begin{array}{ll} \text{one hump} & y = \cos(x) \\ \text{three humps} & y = \cos(3x) \\ \text{five humps} & y = \cos(5x) \end{array}$$

These are called "pure tones."

Now typically, when you pluck a guitar string, you don't produce a pure tone. Instead, you cause the string to vibrate in a complicated way built up out of some combination of the fundamental tone and the overtones. The "main note" that you hear is the fundamental, but the mixture of the overtones adds richness to the sound. The particular combination of fundamental and overtones which you hear produces the characteristic sound of that instrument.

For example, if plucked appropriately, the guitar string might vibrate like this:

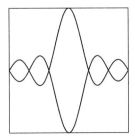

This is a combination of all three of the previous tones. This process of combining different tones to produce a complicated sound, or more generally of combining simple different periodic functions like cosine to produce a complicated periodic function is called *superposition*.

Fourier series are a technique for reversing this process. In other words, Fourier series are a method for taking a complicated waveform or "sound" and splitting it up into a sum of simpler, "pure" tones. For example, we might want to take a "sawtooth" vibration like this:

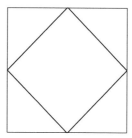

We could then ask what pure tones need to be combined to produce this kind of vibration. The different tones we need to add together are these:

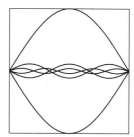

When we add these together, we get a limiting situation like this:

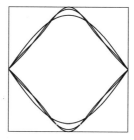

Here you see a series of better and better approximations to the sawtooth made by combining more and more pure tones. It turns out that you need to combine infinitely many different pure tones to do this, and the theory of Fourier series tells you how. Now let's look at some formulas."

Louis slipped from his chair and left the classroom. He was really impressed – and really interested. He was also incredibly embarrassed that he'd never heard even that much explanation of the physical significance of Fourier series. He thought back to what he'd learned in his analysis class – $L^2(S^1)$, Hilbert spaces, all that stuff, which he'd thoroughly mastered – and realized that he had never heard the word "vibration" mentioned!

Right then, he made two resolutions: first, that he'd try to give his class the same motivation Kara had just given him; and second, that he'd go to the library right away and do a little reading on waves.

Friday Tea, One Week Later

"Kara, I want you to know how much I appreciate your having encouraged me to emphasize some of the applications of Fourier series. That little fifteen minute lesson I got from your class opened my eyes to all kinds of interesting stuff. I followed your lead and talked about vibrating strings," Louis said.

"Funny, Louis, I was just thinking that maybe you had the right idea after all. I sort of regret bringing up any of that stuff."

"How come?"

"Several reasons. First, I gave a quiz on Fourier series and I was pretty disappointed in the results. I asked the students to compute a few Fourier coefficients, and got pretty weak responses."

"Yeah, but that's typical, isn't it?" answered Louis.

"Maybe. But I also asked the students on the quiz to describe in their own words something about the significance of Fourier series for understanding vibrating strings. Only a few even tried that question, and of those only two were able to say something reasonable," said Kara.

"I'm not really surprised. It's always hard to get coherent answers to essay questions, especially on quizzes."

"OK. But here's the thing that really pushed me over the edge – when we were talking about the final exam, one of the students asked me if they needed to know any of that "physics stuff" I had covered – not just the

Fourier series, but the extra stuff on work and some other little things I had included as motivation earlier. It kind of put me in a difficult position. I know we're not going to ask any questions about applications on the final, partly because not everyone teaching the course covered that stuff, partly because of tradition, and partly because those kinds of questions are just too hard to grade. But something about the tone of this guy's question made it clear that if I told him that, he would just ignore anything I had said on those subjects and focus on what really matters: the formulas. So why do I bother?"

Louis paused to think. He had found Kara's little lecture on vibration fascinating, and he had naturally assumed that his students would too. But from what Kara was saying, this very material he had found so intriguing was seen by the students as "extra stuff" which got in the way of the "real material." He wondered what his own students would say about his new and exciting (to him) information about vibrating strings. Maybe he should just forget about trying to do anything fancy, and go back to concentrating on the mathematical formalism.

"I don't know what to do, Kara," he finally said, "you may not have done much for your real students, but you helped educate me. I hope that's worth something. Maybe some of it will sink into them, too."

Order Out of Chaos

Terry was excited about today's class. In the pre-Calculus class she was teaching, they had begun considering transformations of functions as a way of building larger families of functions. Terry knew that students typically had difficulty visualizing the graphs of functions, but was confident that she could help them learn some basic principles. The day before she had introduced the idea of translations of graphs, using parabolas as examples, and she was looking forward to reinforcing the concepts today. Terry enjoyed teaching this topic because it gave her a chance to show the geometric meaning behind the algebra.

Terry began class by calling for some of the graphing problems assigned the previous day to be put on the board. Natasha and Ralph volunteered to graph $\dfrac{2}{x}$ and $\dfrac{2}{x-3}$ on $[-2, 2]$.

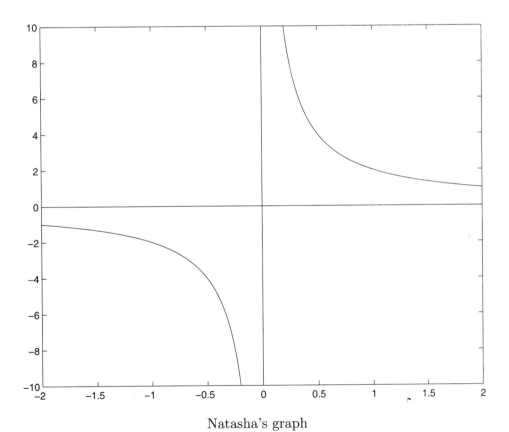

Natasha's graph

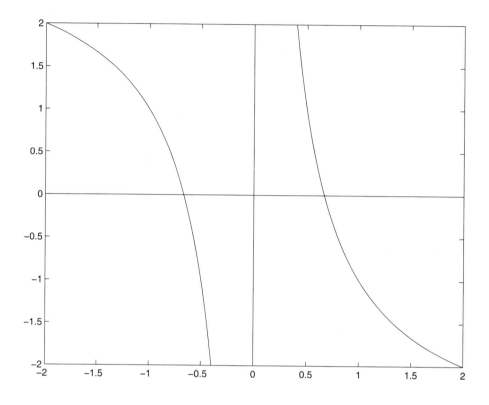

Ralph's graph

Terry looked at the graphs, thought for a few seconds, then turned to the class to ask for comments.

"That's what I got," Bob said. "So did I," added Jane, nodding. Others in the class looked from their papers to the board, saying nothing although some looked puzzled.

Jason announced, "Well I got something different for Ralph's problem!" Terry breathed a silent sigh of relief. "What did you get?" she asked expectantly.

Jason went to the board and confidently sketched a quick graph.

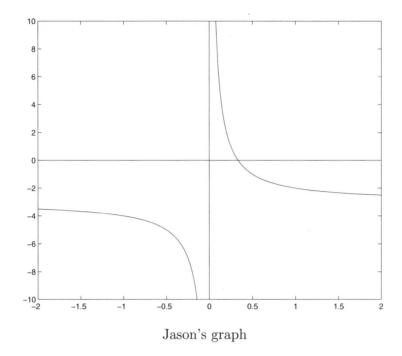

Jason's graph

As she watched him working at the board Terry's spirits began to sink again, but she waited until Jason had finished, all the while wondering what to do next. Her original plan to engage the students about the idea of translations seemed to be going awry.

"Ralph, what do you think?" asked Terry. "What did you do to come up with your graph?"

"Well, I first tried out some points on my calculator ..." Ralph looked at his paper, then put up a short table on the board.

1	-1
2	-2
.5	1
.25	5
.1	17

Ralph continued, "And I was going to go on with this, you know plotting points, when I remembered that my roommate had just showed me how to use the graphing feature on my calculator. So I tried it a couple of times, but all I got was the part on the right and nothing was plotted on the left, so I figured I was just doing something wrong, since there should be some values for when you plug in negative numbers for x. Anyway, I just sort of 'flipped' the graph over to get my picture, kind of like the $\frac{2}{x}$ picture."

Jane and Bob nodded their heads in agreement, with Jane interjecting that while she had made a table with more points than Ralph, she had also sketched a graph based on her table, and she had done a similar "flipping" procedure. Terry noticed that a few others in the class seemed to be agreeing as well.

Jason, who had been trying to get a word in, finally blurted out, "Yeah, well, I messed around with the graphing feature of my calculator too and figured I needed to stretch out the window in the y direction, since you get big values when x is a little number. I mean look at the last couple of pairs of numbers in your table. Anyway, if you do that - I used $[-10, 10]$ for y - you can see the part of the graph that I got for negative values of x."

Ralph exclaimed, "Oh yeah, I forgot about the window settings. There's a lot to keep track of!"

Returning to the board and picking up the chalk, Jason added, "Besides, if you were going to think about 'flipping' the part of the graph on the right for positive values of x, you'd have to kind of do it around a 'negative 45 degree' line like for $\frac{2}{x}$, except it would have to be shifted down some, sort of like this." Jason drew an additional line on his picture.

Terry glanced at her watch and saw that by now nearly 25 minutes had passed. Somewhat worriedly, she turned to the class and asked for further comments about Ralph's and Jason's work.

After some awkward silence, Marilyn quietly volunteered, "I thought for the second problem that it's like really small, when x is really big, like 300 or something. Natasha's graph is like that, but it doesn't seem to match either Ralph's or Jason's pictures." And from the back of the class Alex asked, "Isn't it just the whole thing shifted to the left..., or is to the right?"

Terry thought to herself, "Well, maybe there is a way to salvage this discussion."

Pairing Up

Shalini is a TA for Integral Calculus. She runs discussion sections in which material from lecture is reviewed and practiced. In a typical session, she answers questions for about 10 minutes, then proposes several exercises to be worked by the students during the period. Shalini circulates and helps where needed, encouraging group discussion. She writes on the board only if there is widespread confusion.

Lately, Shalini has felt that the students are relying too heavily on her. She decides to have them work in pairs and explain things to each other, in the hope that it will make them think more for themselves.

Shalini: Today, I want everyone to work with a partner. If you don't know anyone, I'll match you up. You should work on the problem that I'm going to put on the board, and it's really important that both you and your partner always know what's going on. Only ask me questions as a last resort. Questions?

Maria: Do we have to hand it in?

Shalini: No, this is just like any other day, except that you must work together.

She writes the following on the board:

> A cone of height 8 ft and top radius 3 ft is filled with liquid to a depth of 6 ft. Assuming the liquid weighs 10 lb/ft^3, how much work will it take to remove all the liquid out of the top?

Shalini: OK, pair up.

She wanders around for a bit, giving students a chance to settle. Noticing that Evan, a sullen but high achieving student, is about to work alone, Shalini taps on his book and prods, "C'mon, find a partner. Chris, why don't you come work with Evan." Chris is also bright, and more laid back than Evan. Shalini thinks they might complement each other's strengths, and perhaps Evan will recognize the benefits of collaboration.

Shalini is pleased to see that Cassandra and Rae have paired up. Both are quiet, and struggle with the material. Rae often works with her friend Jamie, who is absent today. Jamie has a solid background, and Rae always defers to her, mostly copying Jamie's solutions.

Shalini: How are we doing here?

Rae: OK, I guess.

Cassandra: I don't really get this chapter.

Shalini: Why don't you just start by drawing a picture of the problem? See how much you can label on it. I'll come back in a little while.

Shalini sees Chris and Evan working separately. She goes over, hoping some cheery encouragement will make them communicate.

Shalini: So, where are you two at? Chris, do you agree with what Evan has done so far?

Chris: (guiltily) Uh, we're just working it out first for ourselves, and then we'll compare answers.

Shalini: OK. But today I'd really like everyone to work together. So, Evan, why don't you try and explain your approach to Chris?

Evan: I don't have an approach.

Shalini: That's fine. I just meant—what do you have so far? It looks like a good start. (She sees that he has the basic setup correct.)

Evan: I don't have anything. How do I do this problem? If you would tell me, then I could get it.

Shalini: I think we should start with what you've got. Explain to me what these variables mean? (She points to his equation.)

Evan: Don't you know?

Shalini: I'd like you to tell me.

Evan: You know, I could be this confused at home. I don't need to come here for this.

Chris is frowning at a neat diagram of the problem, but looks up in surprise upon hearing this.

Chris: Gee, I thought you knew how to do all the problems.

Evan: I can do the problems, I just don't understand what they mean.

Chris: Hmm, I have the opposite problem. Let me look at your figure . . . your equation's different from mine . . . oh, I see, you're measuring from the top, not the bottom.

Evan: Yeah, I see . . . it would make the equation easier if I did it the other way around.

Shalini, relieved at not having to respond to Evan's rudeness, leaves them at it and goes across the room to help Trina and Lucas, who have been waving frantically at her.

Trina: (cheerfully) Is this right?

Shalini: Why don't you tell me about it?

Lucas: We split up the cone into pieces, and then integrated.

Shalini: Walk me through it.

Lucas: We made it into circles, and used Work=(force)(distance). Since the force is 10, and the distance is x, we got

$$\int_0^6 10\, x\, dx = 180.$$

Shalini: Trina, do you think it's right?

Trina: It sounds good to me.

Shalini is flustered—she doesn't want to outright say it is wrong, as that will defeat the collaborative approach. But she doesn't want to be the TA whose students have a book full of wrong answers, either.

Shalini: So, you said that you sliced the cone up?

Lucas: Yes.

Shalini: Could you tell me how you used the slices to come up with the integral?

Both students seem confused, and laugh. Trina starts to flip in her textbook.

Shalini: I want you to draw me a slice, and figure out the work to move just that piece. Then see how you can use it in your integral.

Trina: So we're wrong?

Shalini: I didn't say that. I just want us to see where the integral comes from. That way we can tell if it's right.

Shalini leaves Trina and Lucas unsure of what they are looking for, visits some other pairs, and then heads back, as promised, to Cassandra and Rae.

Shalini: How's it going?

Cassandra: We drew a picture. (She points to her page.)

Shalini: OK, what's next.

Rae: Divide it up?

Shalini: Why might we do that?

Rae: Because that's how you always do these problems.

With other students, Shalini might challenge this comment, but she feels Rae's confidence may be too fragile.

Shalini: All right, show me how we should slice it up.

Rae draws some horizontal lines across the cone.

Shalini: Let's draw a typical slice over to the side. (She waits, and when neither student moves, draws one herself.)

Shalini: What is the radius of this slice?

Rae: 3 ft?

Shalini: If I chose a different slice, what would its radius be?

Cassandra's face starts to light up.

Cassandra: It would be different! The radius has to do with where the slice is. It has to do with x! Rae, do you get it?

Suddenly, Shalini sees that people are packing up, and realizes that class is over. She makes a quick announcement, encouraging people to complete the problem, and bring questions next time. She worries that Rae and Cassandra have just begun to understand, but are nowhere near to being able to complete this (or any other) problem on their own. She makes a mental note to talk to Trina and Lucas next time, hoping that they will have discovered their own errors. Shalini cannot decide if she has fulfilled her job as TA today—the only students that she knows will solve the problem are Evan and Chris, and she is not convinced that she has played any part in their success. She hopes that in some capacity she has motivated her students to think independently, but wonders if they will benefit from this skill during the course. The students will need to study this material for the next examination ... perhaps she should just spend the time showing them how to do the problems.

The Quicksand of Problem Four

Background

Bill Baker is a TA assigned to handle discussion sections for Calculus I. The main lectures are given by a professor in a large classroom on Mondays, Wednesdays, and Fridays. Bill's discussion sections review, in smaller groups of 15 or 20, the material covered in lectures. Bill is supposed to make sure the students understand the homework problems, and he has some grading responsibilities as well.

Narrative

Bill briefly goes over in his mind his plans for today's discussion section. This is the fourth discussion section of the semester, and it seems to Bill that things are going reasonably well. He has been trying to conduct the section with plenty of interaction, and in general feels that the students are speaking up and participating in discussions. He knows from the syllabus that the professor talked about instantaneous and average velocity in the lecture the day before, and his plan is to talk about the relationship between these two topics with the students. He especially wants to talk about the nature of the limiting process which relates the two ideas. He figures he'll spend a few minutes answering homework questions, after which he can present some examples he's worked out showing the significance of instantaneous velocity in physical problems.

Entering the room, he begins as planned.

BILL: Yesterday in class the lecturer went over the idea of instantaneous velocity and talked about limits. I figured we could spend today in discussion section looking at some of the problems from this section and reviewing some of these ideas. Before I begin, does anyone have any questions from the lecture?

Bill pauses and looks around the room. No one raises a hand, so he continues speaking.

BILL: OK, so let's start with one of the homework problems. Does anyone have a problem they'd like to look at?

Again Bill looks around the room. He notices some of the students shifting in their seats, but no one says anything or raises their hands.

BILL: Really? Are you sure there are no questions?

35

At this point, one of the students opens his book and speaks up.

JIM: Yeah, can you do problem 3 on page 85?

Bill looks at his book and sees that problem 3 on page 85 is the first assigned problem from the section. The problem is

> *3. Alice travels from Saint Louis to Chicago, a distance of 300 miles. It takes her 5 hours and 45 minutes to cover this distance. What is her average velocity over this time period?*

Bill thinks this seems pretty straightforward. He turns to the class.

BILL: OK, how do you do this problem? Anyone have any suggestions?

He looks at the class and waits what seems to him to be an eternity for someone to volunteer something, but no one speaks.

BILL: Come on, guys, someone here must have looked at this problem...you do know you're supposed to try the homework before coming to discussion section if you want to get anything out of this. Someone here must have some suggestion.

Kathy raises her hand and Bill feels some of the pressure come off.

BILL: Thanks, Kathy, what do you have in mind?

KATHY: Well, you should use the formula for average velocity.

BILL: OK, and what's that?

KATHY (reading from her notebook): It's $(s_1 - s_2)/(t_1 - t_2)$.

BILL: Right! So in this case, we have $s_1 = 0$, that's St. Louis, and $s_2 = 300$, that's Chicago; (he writes on the board: $s_1 = 0$ and $s_2 = 300$); and we have $t_1 = 0$ at St. Louis and $t_2 = 5.75$ at Chicago, (he writes on the board $t_1 = 0$ and $t_2 = 5.75$) and if we use Kathy's formula we get $v_{avg} = (300)/5.75 = 52.17$ miles per hour (he writes $v_{avg} = 300/5.72 = 52.17$ on the board). Does that make sense to everyone?

He looks at the class, making eye contact with Kathy; she nods; Jim, who asked the original question, nods too and starts to write in his notebook. He catches a few other looks from others in the class and decides that he's gotten through that problem without trouble. Now for the next...

BILL: OK, that wasn't too bad. Any more homework problems you'd like to look at?

Jim raises his hand right away.

JIM: How about the next one? Problem 4 on the same page?

Bill figures Jim just wants him to do the homework assignment for him. He remembers his TA training course, where they made a big deal about NOT just doing the problems for the students.

BILL: Well, Jim, you got your question answered last time; how about giving someone else a turn.

Almost immediately a couple of hands go up. Fred speaks out loud:

FRED: I had trouble with Problem 4 too—could you please go over it?

Some of the other students, who have their hands up, make encouraging noises as well. Bill figures he'd better go over the problem—after all, the other thing he remembers from his TA training course is that he should try to respond to the students' interests and difficulties. He decides, though, that he won't just give them the answers; they'll work the problem together.

> *4. Suppose Alice drives the 300 miles from Chicago to St. Louis in 5 hours total driving time. For the first three hours, Alice travels 55 miles per hour. Then she decides she is taking too long to cover the distance. She speeds up to 80 miles per hour for the next hour, when she sees a state patrolman by the side of the road; this makes her slow down and finish the drive at a constant speed.*
>
> *a. Draw a graph showing the distance Alice has traveled as a function of time.*
>
> *b. What is Alice's average velocity for the entire trip?*
>
> *c. Draw a line whose slope shows Alice's average velocity over the entire time.*
>
> *d. Draw another graph showing Alice's instantaneous velocity as a function of time.*

BILL: Let's start out with part (a), where we are supposed to draw a graph of Alice's position as a function of time. Let's start out by looking at the first three hour period. The problem says that Alice travels at 55 miles an hour for these three hours. So what does the graph of position look like for those three hours?

Bill looks around the class. Sarah, who Bill remembers from the past few sections has had reasonable things to say, raises her hand. Before Bill can call on her, though, John interrupts.

JOHN: It's a line.

BILL: Right, it's a line.

Bill turns to the blackboard and puts up a set of axes, labeling the x-axis time (in hours) and the y-axis distance from St. Louis (in miles). Then he turns back to John.

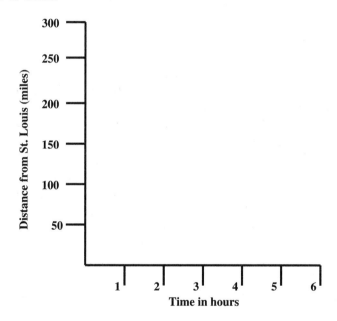

BILL: What line do you have in mind?

JOHN: It's a straight line at height 55.

Bill is surprised at this answer. He is sure the lecturer has reminded the class the day before that motion at a constant speed produces a line whose slope is given by the speed. And anyway, this is pretty basic stuff. He wants to get through this so he can talk about the interesting things. He decides, however, to try to keep working interactively.

BILL: What do other people think. Is that correct?

Sarah raises her hand again, and this time Bill calls on her.

SARAH: It's a line with slope 55 miles per hour.

Bill is relieved. This is more like it.

BILL: Right, it's a line with slope 55. Remember that when travelling at a constant rate, distance is rate times time; and in that situation the graph of distance is a line where the slope gives the rate.

Bill turns to the blackboard and draws in a line with the correct slope. This problem is taking a bit longer than he'd planned—he hadn't really wanted to work through it anyway.

BILL: For the same reason, over the next hour, Alice covers 80 more miles and the graph of her position over the next interval is a line with slope 80.

Bill fills in the next interval. His graph at this point looks like this:

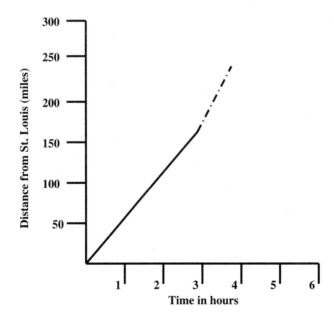

BILL: Now all we need to do is figure out what we need to do to finish the graph and show Alice's arrival in St. Louis. So you see, after 4 hours she's traveled 245 miles; she has to travel 300 miles total; so she has to go another 55 miles. Altogether the trip takes 5 hours. So we can fill in the graph like this:

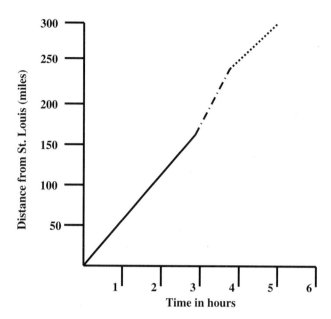

BILL: Finally, you see, we can get her speed over the last interval by looking at the slope of this line—the change in x being 1 hour (he writes $\Delta x = 1$ on the board) and the change in y being 55 miles (he writes $\Delta y = 55$) for a speed of 55 miles per hour. That covers part a, I think.

Bill turns from the board and looks around the class. Several of the students are writing in their notebooks from his description on the board. Most of the others are sitting passively—Bill figures they know this stuff already and are as bored as he feels. He is also starting to feel a little time pressure—all in all, this was taking a lot longer than he thought it would.

BILL: Now part b tells us that we need to determine Alice's average velocity for this entire trip. We've already done one problem like this. So what is her average velocity?

Again there is quiet in the room.

BILL: Remember problem 3 everyone? We already did one computation of average velocity.

After a long pause, Bill sees Jim do something on his calculator, after which he speaks up.

JIM: Her average velocity is 67 miles per hour.

This answer takes Bill completely by surprise. He has been under the impression that the principles behind average velocity are clear to everyone—it is completely straightforward, plus he has already explained it in Problem 3. Trying to keep his voice from sounding testy, he continues.

BILL: Hmmm... Jim, how did you get 67 miles per hour?

JIM: Well, I averaged 55 and 80 miles per hour. They average out to 67 miles per hour.

This surprises Bill even more. It makes so little sense that he is kind of shocked that anyone would think like this—especially since they have heard a lecture the day before in which average velocity is discussed in detail. Before he can get too worked up, though, he decides that he shouldn't panic just because one student made an inane comment. He figures Jim is just not on top of things, and he shouldn't let him undermine Bill's confidence in the other students.

BILL: I think everyone should realize that the use of the term "average" in the phrase "average velocity" doesn't refer to averaging in the usual sense. Kathy, I think it was you who told us how to figure out average velocity in Problem 3. Can you help us out again with this problem?

Bill looks at Kathy, who has been writing steadily in her notebook. She looks up, startled.

KATHY: Well, uh, I'm not sure.

BILL: As I recall, you gave us the formula $v_{\text{avg}} = (s_1 - s_2)/(t_1 - t_2)$. (He writes this formula on the board.) How do we apply this formula in this context?

Kathy shrugs and looks uncomfortable. Bill scans the room seeing a lot of faces carefully schooled to show no interest. Jim avoids his look. His gaze crosses Sarah's and she raises her eyebrows at him—he realizes she understands this completely, but her face offers no sympathy either for him or her fellow students. She keeps quiet.

Bill starts to suspect that most of the students in the class have no idea what is meant by average velocity. He wonders if more than a very few of them have looked at the homework problems at all. His assumption at the beginning of the class that he could quickly review a few homework problems and then move on to a discussion of limits seems hopelessly naive. In addition, he's gotten himself into the middle of this homework problem, and used up 25 minutes of his 50 minutes, and it looks like it's all been wasted. He starts to feel very depressed, and with a sinking feeling in his stomach, he tries to decide what to do....

Salad Days

Evan shuffled nervously into the hall outside his TA's office in the basement. He was there to ask for an extension on a lab report his calc I class was supposed to write up, making conjectures about the effect of parameters on graphs he was asked to draw using Mathematica. He had begun the assignment. His major was agronomy, and unlike most of his engineer classmates, he did not already know how to use the software; computers in general were awkward territory for him. He'd spent ten hours in the computer lab, learning how to use the program, but felt it would take another ten to complete the project. He was committed to participate in a mentoring internship program with the Future Farmers of America all weekend during the hours the lab would be open. His earlier attempts to visit the TA that week had been unintentionally thwarted by his roommate, Rick, who was deeply depressed, not handling the transition to university well, and needed, Evan felt, constant attention.

Exactly at three, Arnold, the TA, opened his door. "Come in Evan, what can I do for you?"

"Ah, well, I was hoping to ask you about the project. It's due next Monday?"

"Yes, that's right; we want a write up, with print-outs of some pictures you came up with, a conjecture about what happens for different values of n, and an explanation in clear English of how you arrived at your conjecture." Arnold was very pleased with this assignment; it had been his idea, and he thought it would give the students a real taste of mathematical exploration, not the usual plug-and-chug of the calculus sequence. He looked forward to seeing what people had done.

"Yeah, I was wondering if it would be possible to turn it in a little late, like say, on Thursday... I've never used Mathematica before, and I have to work this weekend."

Arnold, frowned, then said, "The grade goes down by a third of a letter grade for every day it is late. So an A paper received on Thursday would get a B. If you are willing to see your grade lowered like that, go ahead and hand it in Thursday."

Two weeks later, Arnold sat in his office, grading papers. He came upon Evan's project, not three but five days late. He had done a thorough job, and his comments were well-written, but the conjecture was slightly inaccurate. If it had been turned in on time, he would have given it a B+. Five days late, it was a C-. As he entered the grades in the grade book, he saw that that

43

Evan had received a low C on the first exam, and a high D on the second one, taken a week after the project was due. "That kid is going to end up taking this course again if he's not careful," Arnold thought to himself.

A month after that, Evan was back in office hours, this time studying for the third exam. As the TA patiently explained Riemann sums for the third time, Evan's head started to swim. He had been up all night helping one of his FFA advisee's sows farrow. Having lost track of the notation, he gave up listening entirely, and started calculating in his head how well he would have to do on this test and the final to pass the course with a C+, the minimum grade required for his major. He spoke up.

"So I need to get at least a B- on this test to bring my average up to a C+ before the final, right?"

Suppressing a sigh, Arnold finished explaining the problem to the other students there. After they left, he turned to Evan.

"If I remember correctly, you have a borderline D+ right now, although I remind you that this is only approximate; the grades will be averaged and curved at the end of the semester. In any case, you should certainly try to do the best you can. I would say that anything less than a B is cause for concern. I'm sorry, but I have another class now, so I have to go. Perhaps we can talk about this after class tomorrow."

Dazed, Evan heaved his backpack out into the hallway. The TA made it sound like he didn't think Evan had much chance at all of making it as things stood at the moment. He wished he had waited to take this course in the spring, when his internship was less demanding. He wondered if he could get out of the course somehow, and if that would affect his financial aid. He wondered what impact failing the course would have.

Seeking Points

Daniel sighed as he dumped his books on his office desk. He'd just handed back the first midterm exam from his Calculus I class, and he could tell as he left the classroom that there were a lot of unhappy students. Still, the exam had been just like the practice exam he'd given out, and he was sure it was pretty straightforward. As he sat down to take a look at the paper on duality for fppf sheaves he was supposed to read, he heard a knock on his office door.

"Come in," he called, and he saw Sam, one of his Calculus students, push open the door hesitantly.

"Can I talk to you about my exam?" Sam said.

"I guess this was inevitable," thought Daniel to himself. To Sam, he said "What's up?"

"It's this question number 2," said Sam. "I don't think my answer was graded properly."

"Let me take a look," Daniel replied, "pull up a chair."

Sam sat down and passed his exam booklet over to Daniel. Daniel noticed that Sam had gotten 82 points out of 100 on the exam, which was a high B; but he had missed most of his points on problem 2. Then Daniel looked at the question, which said:

Problem 2 (20 points). Let $f(x) = x^3 - 5$. Use the definition of the derivative to compute the slope of the tangent line to the graph of $f(x)$ at the point where $x = 2$.

Then Daniel turned to Sam's exam paper. Sam had written the following:

Sam's Answer: $f(x) = x^3 - 5$. $f'(x) = 3x^2$. Slope=$f'(2) = 12$.

The grader of the problem had given Sam 5 out of the 20 points.

"Well, Sam," said Daniel, "you see you didn't do what the question asked. You are supposed to use the definition of the derivative to solve this problem, but you didn't give any method for deriving your answer. How did you do this problem?"

"I used the rule that the derivative of x^n is nx^{n-1}, which makes it really easy," replied Sam.

Daniel felt a little uncomfortable about this. He, like the rest of the Calculus teachers, was emphasizing understanding rather than algorithms for solving problems. He and his fellow instructors had specifically scheduled the first exam after a qualitative discussion of the derivative, and an introduction to the definition, but before discussing the various techniques of differentiation. He hadn't gone over the nx^{n-1} rule in class yet.

"Where did you get that from?" he asked Sam.

"I took Calc in High School, and we learned it there. We learned lots of other methods too. The answer is right, isn't it?"

"Yes, it's correct as far as it goes, but as I said it isn't what we asked for. We wanted you to show that you can use the definition of the derivative."

"You mean that thing with the limit?" said Sam.

"Yes," said Daniel, "Exactly, that thing with the limit – the difference quotient. In the review for the test I emphasized that if we asked you to use the definition of the derivative then we wanted you to use the difference quotient."

"Well," said Sam, "I didn't come to the review session. But it doesn't really seem fair to me that I got so many points off because I did the problem an easy way instead of a hard way."

"It isn't just a question of easy and hard," said Daniel. "We are trying to teach you to understand what the derivative means and where it comes from. We don't want you to just learn a bunch of formulas and how to make them go."

"Look, Professor, I know what the derivative means. It's the slope of the tangent line to the curve at the point, just like you asked. I knew that, because I knew what to calculate once I used my rule. Look at the rest of my exam – I got all the other problems basically right. I think I deserve more points on this problem."

"Sam, before we get into a discussion of points, let me ask you this. Do you know what the difference quotient is? Do you know WHY the formula you used gives you the slope of the tangent line?"

"Yeah, well, you did that in class a while ago, and I understood it then. It has something to do with secant lines and stuff, but I forget right now. I figured it doesn't really matter, 'cause I know these other, easier ways to do the problems. I just feel sorry for the other students who have to do it the hard way. I taught my roommate in another section about the methods I learned and he really appreciated it."

"But Sam, that's just the point we are trying to get across. It IS just as important to know WHY the formula works as how to use it. The formulas you learned all had to be figured out by someone using the difference quotient. Let's take the problem from the test. What we wanted to see was the following:

$$
\begin{aligned}
f'(2) &= \lim_{h \to 0} \frac{((2+h)^3 - 5) - (2^3 - 5)}{h} \\
&= \lim_{h \to 0} \frac{2^3 + 3 \cdot 2^2 \cdot h + 3 \cdot 2 \cdot h^2 + h^3 - 5 - 2^3 + 5}{h} \\
&= \lim_{h \to 0} \frac{12h + 6h^2 + h^3}{h} \\
&= \lim_{h \to 0} 12 + 6h + h^2 \\
&= 12
\end{aligned}
$$

So the derivative is 12, and so is the slope. This calculation shows that the slope, 12, is the limiting value of the slopes of the secant lines."

"Well, maybe it shows that to you, but it looks like a bunch of formulas to me. Just different formulas. You really think all those people who wrote that instead of what I wrote know something I don't? They just went to the review session, which I admit I shoulda done. Look, Prof, I'm not here to argue about all of Mathematics. I promise from now on I'll come to your review sessions and do the problems just the way you want them. I just want 5 more points so I can get an A on this exam."

Daniel's heart sank. It was pretty clear this kid Sam didn't get Daniel's argument about "underlying ideas." And him promising to do whatever Daniel wanted on the next exam just made Daniel fell worse – that made it seem like the kid was just humoring him. As for more points – well, lots of people had made this mistake on the exam, and they'd all gotten five points. So Daniel couldn't really change this kid's point score without changing the others, too, though he did think Sam seemed pretty sharp.

"Sam, I'm afraid I can't give you any more points on this problem. We graded the exam consistently, and we gave everyone who made your mistake 5 points. I appreciate what you're telling me, and I get the impression you are following the course pretty well, so if you continue to do well you can get your A on the next midterm and the final and you'll get your A in the course."

"So you mean you graded lots of people unfairly, and you don't want to fix it? OK, you're the prof, I guess. And I'll be sure to come to the review session next time so I find out how you want us to do the problems."

Sam picked up his exam and left the room. Daniel stared after him for a minute or two, visibly upset, then took a deep breath and turned back to his desk. He had promised to read this paper before his next meeting with his advisor. Where was he? Oh, yes, he could see that the argument he was reading worked if he used the theorem on flat descent. He remembered sitting in on a lecture during his second year where his professor had described flat descent in detail; he couldn't exactly remember the proof of the theorem, but he did remember that you could apply it in this situation....

Study Habits

Angelica: OK, that is the end of our class for today. Be sure to review Chapter 6, especially the discussion of integration techniques. As usual, the recommended homework assignments are listed on the syllabus. You all need to keep up with the homework. See you later....

Angelica collected her notes and sighed quietly to herself. She had been leading her section of second semester Calculus for six weeks. She liked being the "professor", instead of just working as a grader or assistant for some faculty member; it was nice to work with relative independence. True, the syllabus was laid out in detail, and the hour tests and final exam were to be given in common to all of the sections of the course; but the day-to-day classroom work was up to her.

The fact that she had so much independence made it particularly painful when she thought about how poorly her students had done on the first test – no worse than anyone else's really, but still it had been disappointing, since she had worked so hard.

Walking back to her office after class, she bumped into Simon and Kathy, two other graduate students working on the same course.

Simon: Hello Angelica; you look glum. What's the problem?

Angelica: Nothing really; but I was really looking forward to teaching this course and now I am finding it kind of depressing. The students are so passive, and most of them are so unprepared. Do you think they study at all?

Simon: I am not sure, Angelica. But attitudes about school certainly seem different here in America than they are in my country. In order for me to go to university, I had to pass many very difficult exams. Then when I was accepted I felt very privileged. In fact, we students worked very hard all the time because if we didn't do so, we could be dismissed. I have learned not to expect the same behavior from my students here. I review the material in the book, I answer questions if there are any, I go over problems or talk about what I think are the hard parts of the course – and I try not to worry about what the students are actually learning.

Angelica: I also had to work very hard at school in my country and had to pass many exams to get into university, and to stay there. I feel that our students don't work hard enough. They expect us to make everything easy for them.

Kathy: I realize both of you worked very hard to get into university, but you shouldn't think our students didn't do likewise. I know lots of them are the first ones in their families to go to college. In addition, many are holding down full time jobs to pay for their education.

Angelica: That may be true, but if they are here in school then their schoolwork has to come first. I have thought about this, and I think that we have to force our students to work harder. I think we need more than just hour tests and a final exam. I think we need to have many more graded assignments, like graded homework and quizzes. If we force the students to work more between exams, they will change their priorities.

Simon: I don't think you can make a change like this on your own. This class is being run as a group, and everyone has to do the same things in class.

Kathy: Besides, Angelica, are you sure this is a good idea? We'll have to grade all of those extra assignments you are proposing. It's going to mean a lot more work for us, too. Perhaps you should leave well enough alone.

Angelica: I am willing to do the extra work. I think it is necessary. I will talk to Professor Jacobs about my idea of having much more graded work.

The three students split up to go to their other classes. Angelica dropped by Professor Jacobs's office later that day with her plan. Professor Jacobs was on the phone when she knocked on his door, but he motioned her into the office to wait while he finished his conversation.

Angelica: Hello Professor Jacobs. Can I speak to you about Calculus?

PJ: Sure, Angelica. What's up?

Angelica: I want to ask you a question about Calc II, since you are the course coordinator.

PJ: OK.

Angelica: You see, Professor, my students didn't do very well on the first test, and –

PJ: (interrupting) I don't think they did any worse than the other sections, did they?

Angelica: No, about the same I think. But still, I was disappointed – so many of them seemed to miss problems which I know I explained in class. I wanted to ask you about a plan I have which might help.

PJ: What do you have in mind?

At this point, Professor Jacobs's phone rang, and he answered it. Angelica sat patiently while he discussed possible meeting times for a committee he was chairing. After a few minutes, he hung up and turned back to her.

PJ: Sorry about that. You were saying?

Angelica: I think the problem is that the students don't study enough. I think if I start a system of giving them frequent quizzes and also grading their homework assignments, I can make them keep up with the class.

PJ: You mean you want the quizzes to count on their final grade? I don't think that would be fair. I mean, for you to do this, but not the other sections.

Angelica: I think all the sections should do this. By having many more graded assignments, we will prevent the students from falling behind. They will learn the material better.

PJ: Angelica, I appreciate your concern for your students. But I see some practical problems with your proposal. For one thing, this will make an enormous amount of work for you and the other section leaders. You'll have to grade all of these extra assignments, after all.

Angelica: I am willing to do this, if it will make the students learn more.

PJ: I worry that spending so much time on grading will interfere with your progress in your own education.

Angelica: I am sure that I can find the time I need, and I think the other graduate students can too.

PJ: I don't know if your friends will agree with you, Angelica. We expect you to spend a reasonable amount of time on your teaching, but we don't want you to overdo it. There is another problem with your proposal. Do you think it is fair to change the grading scheme in the course in the middle? We told the students at the beginning of the semester how we would grade them, and we owe it to them to stick to what we said.

Angelica: But Professor Jacobs, I don't think it is working. If they aren't learning, shouldn't we do what is best to help them?

PJ: I have a suggestion. Why not talk to your students and see what they think?

Angelica: Aren't we supposed to know what is the best way to teach them? Is it really necessary to talk to them?

PJ: Angelica, I'm willing to consider your proposal, but I think you need to hear from the students first.

Angelica: All right, I will talk to them.

The next day, in class, Angelica nervously brought up the subject.

Angelica: Class, I wanted to speak to you today about the way we have been working. Many people in this section did poorly on the test and I have noticed in our class discussions that many of you do not seem to have read the book or done any homework in preparation for our sections. I think that your lack of preparation for class is a serious problem and I would like to do something about it. I have thought about my own experiences learning mathematics, and I know that I had to work hard at it. I have suggested to the course coordinator that we begin a regular system of brief quizzes, at least twice a week, and that your scores on these quizzes would count towards your final grade. The course coordinator wanted me to ask you what you thought of this idea. Do any of you have any comments?

There was a long silence, but finally Sam, one of Angelica's weaker students, spoke up.

Sam: I don't think that's fair. You shouldn't change the grading in the middle of the class.

Angelica: But we would be doing this to help you. And everyone would have the same quizzes, so it would still be fair; you would just have to try harder to keep up.

Sam: I don't have the time to do so much math every day. I have other courses, too.

Angelica: But that is what I am saying – you must spend more time on math if you want to learn it.

Sam: My goal is to get a C in this course. I need the math credits, but I don't care about anything else. These quizzes just make it harder for me.

At this point, Kim spoke up. Kim had performed well on the last test, and Angelica had always thought of her as one of the better students.

Kim: And what about if we miss a quiz? How would you count that?

Angelica: If you miss a quiz, you get a failing grade on that quiz. The whole point is to make you come to class and do the work.

Kim: I don't think that's fair. For one thing, I have to work, and sometimes I need to miss class for my job. And for another thing, none of my other courses do stuff like this to make me come to class. You shouldn't try to run our lives like this. This is supposed to be college. It's our responsibility to learn stuff.

Angelica felt a bit flustered. Kim seemed angry, as if Angelica had insulted her. Before she could think of what to say, Fred and Bill joined the discussion.

Fred: Angelica, I think having quizzes is a good idea. I don't like the way my grade is based on just a few things. It doesn't leave much room for me to mess up.

Bill: Would every class take the same quizzes? Because I worry that your quizzes would be harder than the other sections, and that would make our grades lower.

Fred: And besides, I like having lots of quizzes because then I see the kinds of problems you ask on tests. That helps me study for the tests, because I know which problems to study and which aren't important.

Angelica wasn't sure why Professor Jacobs wanted her to talk to the students – each of them seemed to have their own opinion on the subject, and none of them agreed.

Angelica: I see that people feel differently about this idea. Let us go on with class now. I will talk to Professor Jacobs again.

After class, as Angelica left the classroom, she overheard two of her students, John and Sally, talking in the hallway.

John: Maybe if she took the time to explain things so we could understand them, we'd do better on the tests. It's typical that she thinks it's our fault. Bad teachers always blame their students.

Sally: Oh, she's not so bad. I've had worse. Why, last semester....

Sally's voice trailed off as she walked away from Angelica down the hall. Angelica felt crushed, and kind of angry. Walking back to her office, she thought over her plan for quizzes. It was a good idea – she was sure of it – but why make extra work for herself, when her own students did not seem to care anyway? And Professor Jacobs had seen only the obstacles to the problem. She stopped by his office later that day.

Angelica: Professor Jacobs, I have decided that my idea about quizzes and grades won't work. So don't worry about it anymore.

PJ: Angelica, I'm pleased that you've talked with your students. Be aware we've tried stuff like that before, and I've never noticed it to make much difference. But let's discuss your concerns more at another time. I have to go to class now.

As she walked down the hall, Angelica thought about the whole experience. No, there was no point in trying to make the students work harder, at least not now. She would do what Simon had suggested: cover the material, and try not to feel bad. Maybe someday she would have good students, and the freedom to teach the way she wanted to....

Studying the Exam

The following questions have been proposed for a 50-minute midterm. Select the problems that you would use for the exam, giving reasons.

If you feel that there are some important gaps in the problems given, feel free to include problems of your own. If some of the problems need re-wording, feel free to do that also.

In constructing the exam, you should consider the following questions, as well as any others you think important.

- Is the exam the right length?

- Do the problems work out cleanly enough?

- Which questions will be hard to grade? What advice would you give to the graders of each problem?

- Is the exam balanced? In topics? In style of problem?

- Are the instructions to each question clear? Are there any that you think students could misinterpret?

- If the students are using calculators, how will this affect the way in which they approach each problem?

- What order do you think the problems you choose should be in? Do you think it matters?

- How will students do who know only the basics? Will they get some points? Few?

- How will students do who have worked hard and understood most topics, but not all?

- Which problems, if any, rely on students having got the first step right to be able to attempt the main part of the problem? What do you think about such problems?

- Should the students be given a choice of problems on the midterm?

Studying the Exam

College Algebra Questions

These questions are suggested for an exam on linear and exponential functions. Your discussion leader will let you know what technology (for example, graphing calculators) is available.

1. Solve for x: $2^x = 3$.

2. Solve exactly: $5^{2x-1} = 0.2$.

3. Find the value of x: $4^x - 2^{x+1} + 35 = 0$.

4. Find the equation of the line through the points $(5, 1)$ and $(-2, 3)$.

5. What is the equation of the line perpendicular to the line $y = ax + b$ and through the point $(2, 3)$?

6. Find the equation of the line parallel to the line $ay + bx = c$, but with double the x-intercept.

7. Graph $y = x^2$, $y = (x - 1)^2$, $y = x^2 - 1$, and $x = y^2 - 1$.

8. In 1996, about 80 out of 1000 obstetricians were sued for malpractice; in 1998, about 100 out of 1000 were sued. Assume that the percentage of obstetricians sued is a linear function of time, t, in years since 1996. Write a function giving the percentage of obstetricians sued as a function of t. (Source: RMF Quarterly, Spring 2000, newsletter of the Risk Management Foundation, Harvard Medical Institutions, 101 Main Street, Cambridge MA 02142.)

9. A ship is traveling at 20 mph along the line $y = 2x + 10$, starting at the point $(-10, -10)$ when $t = 0$. A small boat starts at the point $(0, 40)$, and travels along the line $y = 40 - x$ at 20 mph. Do the boats collide? If so, when and where?

10. Are the following statements true or false?

(a)
$$\frac{4a^0}{(12a)^{-1}} = 48a.$$

(b)

$$\sqrt{4p + 4q} \cdot \frac{1}{(q+p)^2} = 4(q+p)^{-2.5}$$

(c)

$$2x^2 + \frac{3+x}{x^2} = 5 + x$$

(d)

$$\log(A^3 + B^2) = 3\log A + 2\log B$$

Studying the Exam

Calculus II Questions

The following problems are suggested for a Calculus II exam on techniques of integration and applications of definite integrals to volumes and motion. Calculators with a computer algebra system are not allowed. Your discussion leader will let you know what technology (if any) students will have available.

1. Find

$$\int 2^\alpha \cos(3\alpha)\, d\alpha.$$

2. Find

$$\int \theta\, e^\theta\, d\theta.$$

3. Find

$$\int 3x^2(1+x^3)^9\, dx.$$

4. Find

$$\int t^3 e^{t^2}\, dt.$$

5. Find

$$\int \sin^3 t\, \cos t\, dt.$$

6. (a) Calculate $\int_0^2 x^2 - x\, dx$.
(b) Find the area between the curve $y = x^2 - x$, the lines $x = 0$ and $x = 2$, and the x-axis.

7. Find the volume of the region obtained by revolving the region $y = x^2 - x$, with $0 \le x \le 2$, around the x-axis. Sketch the region and show your reasoning.

8. Find the volume of the region whose base is the disc $x^2 + y^2 \le 1$, and whose cross-sections perpendicular to the x-axis are squares.

9. A particle moves along the curve defined by the equation $y = x^3 - 3x$. The x-coordinate of the particle, $x(t)$, satisfies the equation

$$\frac{dx}{dt} = \frac{1}{\sqrt{2t+1}} \quad \text{for } t \geq 0$$

with initial condition $x(0) = -4$.

    (a) Find $x(t)$ in terms of t.

    (b) Find

$$\frac{dy}{dt}$$

    in terms of t.

    (c) Find the location and speed of the particle at time $t = 4$.

(This is a 1998 Advanced Placement$^{\circledR}$ examination problem. Reprinted by permission of the College Entrance Examination Board, the copyright owner.)

10. A particle moves along the curve $y = x^3 - 3x$. The x-coordinate of the particle satisfies the equation

$$\frac{dx}{dt} = \frac{1}{\sqrt{2t+1}} \quad \text{for } t \geq 0$$

and $x = -4$ when $t = 0$. Find the location and speed of the particle when $t = 4$.

11. (a) Show that the total distance traveled by a car moving with velocity, for $t \geq 0$,

$$v(t) = \frac{50}{e^{(\ln 60)t}} \quad \text{miles per hour}$$

is, in miles,

$$\int_0^\infty \frac{50}{e^{(\ln 60)t}} \, dt.$$

    (b) For $t \geq 0$, does the car travel less than 15 miles in total?

Studying the Exam

Multivariable Calculus Questions

These questions are suggested for a midterm covering partial derivatives, either with or without graphing calculators.

1. Find f_x if

$$f(x, y) = x^2 y^3 + 6x^2 + 7y.$$

2. Find

$$\frac{\partial P}{\partial \ell},$$

if

$$P = 2\pi \sqrt{\frac{\ell}{g}}.$$

3. Find g_y if

$$g(x, y) = e^{\sin(x^2 + y^2)}$$

4. Given $f(x, y) = x^2 + y^3 + xy^2$,

   (a) Find $f_x(1, 2)$ and $f_y(1, 2)$.

   (b) Estimate $f(1.1, 1.98)$.

5. Given a differentiable function, $f(x, y)$, with $f(3, 4) = 5$, $f_x(3, 4) = 3$, $f_y(3, 4) = -2$, estimate $f(2.9, 4.02)$.

6. Find the rate of change of z with respect to c if

$$z = 2^{\tan\left(xc^2 + \sqrt{c + \sqrt{c}}\right)}.$$

7. Calculate grad f at $(1, 2)$, if $f(x, y) = 3x^3 + xe^y$.

8. Find the rate of change of g at the point $(3, 4)$ in the direction of $(1, 2)$ if

$$g(x, y) = x^2 y^3 + xy^2.$$

9. Are the following quantities vectors or scalars?

(a)
$$\frac{\partial f}{\partial x}(a, b),$$

where $f(x, y) = \sin(xy^2)$.

(b) The directional derivative of $f(x, y) = xy^2 + e^x$ at the point $(1, 2)$ in the direction $\vec{u} = 0.6\vec{i} + 0.8\vec{j}$.

(c) The gradient of $f(x, y) = \cos(x + y^2)$, evaluated at $(\pi, 0)$.

(d) $(\mathrm{grad} f \cdot \vec{u})\vec{u}$.

10. Suppose $w = f(x, y, z) = xe^{y^2} + z^2 \cos(x + y)$ and $x = \sin t$, $y = t + 1$, $z = t^2$. Calculate $\frac{dw}{dt}$.

11. Suppose $w = f(x, y)$ and $x = g(u, v)$, $y = h(u, v)$. If $f(3, 4) = 5$, $g(1, 2) = 3$, and $h(1, 2) = 4$, find $f_u(1, 2)$.

12. The temperature at the point (x, y, z) is given by the function

$$T(x, y, z) = x^2 + 2y^3 + 3z^4.$$

A bug at the point $(2, 1, 2)$ moves in such a way as to increase the temperature fastest.

(a) In which direction does the bug move?

(b) Suppose that x, y, z are measured in centimeters, T is measured in °Celsius, and the bug moves at 2 cm/sec. How fast, in °Celsius per second, is its temperature changing as it leaves the point $(2, 1, 2)$?

13. Maximize $x^2 y^2 z^2$ on the surface $x^2 + y^2 + z^2 = 1$.

There's Something about Ted

Part I

Ted wasn't sure, but he thought he was in big trouble. He stood nervously waiting to meet with his department chair, Professor Wilson, outside Wilson's office. He wasn't sure why he'd been summoned, but he had a feeling this had something to do with his Calc class. He'd started teaching with such high hopes, but this class had somehow turned into an ongoing nightmare.

Wilson: Hi Ted, come on in. Thanks for stopping by. Have a seat. There are a few things I need to discuss with you.

Ted: Sure, Professor Wilson.

Wilson: You can call me Jim, Ted – now that you're a faculty member, I think we can drop the titles, OK?

Ted: OK, Jim.

Wilson: Listen, I called you in here because I've had a number of student complaints about your Calculus section. I must say that this isn't that unusual – students often complain – but in your case the complaints are a bit stronger and a bit more numerous than the usual background noise. How is your class going?

Ted: Well, uh, I think it's going OK.

Ted grimaced to himself. I was right, he thought, this is about my class. This nightmare is just getting worse. The irritating thing is that in grad school at HSU I was a popular teacher. Of course, *that* school had serious entrance requirements and the students knew how to work. Here, at OU, they seem to admit anyone who is old enough. Half my students never show up in class, and they sure don't seem to care about anything I have to say. Plus they do horribly on the exams, and talk in class, and now they're complaining to the department chair about me!

Wilson: Well, I must say I'm a bit concerned, based on these student comments. The department takes teaching very seriously, Ted, as does the higher administration. You come highly recommended to us as a teacher as well as an outstanding researcher. Why don't you see what you can do to address the situation, and then let me know how things are going?

Ted: OK, I can do that.

Ted walked quickly out of the office, with his stomach in knots, thinking about his course. He felt strongly that Calculus needed to be presented as a serious subject, and he had been relying on the same notes he'd taught from at HSU. The students at OU were entitled to just as complete a Calculus course as the HSU students had been, and Ted had been determined to give it to them. Everyone at HSU had thought his lectures were models of clarity. However, based on how students here reacted, it seemed that he would have to change something, especially since it appeared that the department chair was going to take their opinions seriously. The question was, what should he do now.

What Were They Thinking?

Hugh walked slowly back to his office, puzzled by his class's reaction to the midterm. Since he had been careful to cover all the material and in plenty of time for students to review, he thought that they would find it pretty straightforward. However, from their faces and comments as they handed the exams in it sounded like either he, or they, had misjudged:

> *Man, that was hard. How'd you expect us to do that?*

> *That was nothing—I mean nothing—like the homework.*

> *I studied so hard, but it didn't help in the slightest on those problems. Next time, I may as well wing it for all the help studying gave me.*

Hugh had planned to grade the exams that Friday afternoon, before digging in for the weekend to get ready for his first presentation at his new advisor's seminar. But now he was not looking forward to looking at the exams.

Hugh was in his second year of graduate school. He had found the first year of graduate school hard, but not impossible. Sometimes he felt as though he was surrounded by people who knew what they wanted to work on and whom they wanted to work with. He was envious of the foreign students who had done so much more than he before they entered graduate school. But, he had reminded himself, he would have an easier time adjusting to teaching in an American classroom than they.

Over the summer, Hugh read a great deal and eventually gathered enough courage to make an appointment to ask Professor Zwecki to be his advisor. Igor Zwecki, who was internationally known, had recently moved to Chestnut Valley University as the university had been willing to accommodate his many graduate students. Professor Zwecki looked surprised to see Hugh, but agreed to take him on if Hugh could keep up with the weekly seminar that he insisted all his grad students attend. Hugh, not feeling as confident as he sounded, said that he was sure he could.

Although his own course work had sometimes been grueling, Hugh had found teaching last year to be relatively easy. As a recitation instructor in a large lecture section of Calculus I he had enjoyed explaining the points that he felt the professor had skipped over, and he had found that the students were happy provided that he went over all of their homework questions. They didn't come to office hours much, but that didn't surprise him, as he had not gone to office hours himself as an undergraduate—it was almost a matter of pride that he should figure out the answers to his own questions. He had several students transfer to his section, saying that they had found him much easier to understand than the other TAs. The professor had a few meetings with the TAs, but largely for administrative purposes, so Hugh's assessment of his own teaching was based on the end-of-the-semester evaluations. These were not strong for the course as a whole:

> *It would help if the professor would explain how to do the problems before they are due, not after.*

> *Book sucks. It is only useful for someone who has had the course before.*

> *How about giving us enough time to do the exams?*

However, the comments on Hugh's performance as a TA were encouraging:

> *Brightman was the best TA.*

> *Mostly clear, and could do all of the homework problems.*

> *I wouldn't have got through the course if it wasn't for the notes that I got in Brightman's sections.*

Hugh had expected to be selected to be one of the TAs who taught their own course under faculty supervision as early as second year. He had misgivings about doing this since he knew that he needed to concentrate on getting his own work moving. When he was offered the chance to teach his own course he had talked it over with Paul, his officemate. Paul had said that it wasn't necessarily too much work if Hugh were to arrange to teach in a course where the supervisor did not hold too many meetings. Paul told Hugh whom to request and said that he could certainly borrow Paul's old exams and lecture notes. Of course, Paul had pointed out, the course might be a bit different now, since the department had switched books. But he didn't think that would really make much difference. Grateful for the advice, Hugh had requested, and got, a section of Calculus II under Professor Gatewood. Mike Gatewood was an experienced and popular teacher who had got the process of running a large course down to a fine art. Efficiency was particularly on his mind at the moment as he was in the running for the open deanship. So far the term was exactly as Hugh had hoped—few course meetings and no major problems.

Well before the first exam Hugh went through Paul's old exams and chose several of the more interesting problems from them. He added a few of his own and then showed it to Paul, who suggested that the last one might be too hard. Hugh cut it down a bit but pointed out to Paul that he had done one just like it in class. It was a problem on exponential decay in which the half-life was given; the question was to calculate the fraction remaining after two half-lives. Hugh had asked for the fraction remaining after n half-lives too, but omitted that after talking to Paul. He went by Professor Gatewood's office several times, hoping to show him the exam. The only time the office door was not shut he saw Gatewood in the middle of what looked like a long phone call, and so decided to leave the exam in Gatewood's mailbox. Later that day the exam was returned with two typos marked but no other comments.

Hugh's students were half freshmen, fresh from high school calculus, and half upperclassmen. In class they were quiet and seemed bored. Hugh knew that for many of them integration was review, and several times apologized for going over material that most people knew. But, he said, he was required to go over it again as not everyone had had it. He tried to ask questions but no one seemed inclined to answer, so he decided not to push it. As with the office hours, Hugh remembered how he had felt as an undergraduate, when he often preferred just to hear the lecture uninterrupted by questions and diversions.

Hugh decided to make the material more interesting by explaining the underpinnings—a bit on existence and uniqueness as he started differential equations, and he was planning on including something on Fourier series to liven up the section on series. The homework problems he graded were, in general, fine. He was surprised when so few students came to ask questions before the exam, but convinced himself that they really were finding the course too easy.

So nothing led Hugh to expect the directness and near venom that he felt when students were handing in the exam. He decided to go to the library and see how bad they really were. As soon as he started to grade his heart sank. He couldn't believe that they couldn't integrate $x^2 \cos x$ and that several papers thought that the integral of $1/x^2$ was $\ln x^2$. How could they possibly think this, Hugh thought—I spent so much time making sure they saw the definition of the natural log as the integral of $1/x$.

With deepening concern, Hugh turned to the exponential decay problem. It was a disaster. How could they possibly not have realized that half-life means half the stuff decays, and so in two half-lives one quarter remains? But instead of the two-line answer Hugh had expected, he was faced with pages and pages of messy calculations involving logarithms (often manipulated by rules he had never seen before) and trailing off into messy answers to six decimal places. What *were* they thinking?

Part II

Supporting Materials for Faculty

Developing Effective Mathematics Teaching Assistants
Using Case Studies: An Introduction for Faculty

1. Overview

Case studies are fictional scenarios depicting life-like situations, to be analyzed and discussed. The use of case studies as a teaching tool is common in certain disciplines, such as law and business; cases are also used in many subjects for faculty development at both the university and pre-collegiate levels [1]. They allow the presentation of complicated classroom situations from multiple perspectives, and the discussion of aspects of teaching which go beyond the mechanical. The Boston College Mathematics Case Studies Project, BCCase, has developed a set of case studies intended for mathematics graduate students. This note is intended to introduce faculty members to using them with such students. It is a supplement to the overall introduction to case studies given on the previous pages.

The case studies we have developed are university-level mathematics classroom scenarios, to be read and discussed. They are meant to be used in a TA-development setting, such as a TA-development seminar which meets periodically throughout the academic year. Each case will typically be used for one or two training periods (one hour to one hour and a half each), though the list of questions provided in each Teaching Guide is extensive enough for more time or more sessions. Participants—which may include advanced TAs or faculty as well as beginning TAs—read the case [2]. This is followed by an analytical discussion, facilitated by a faculty member.

The goals of the discussion are both the development of insights for dealing with the particular problems at hand and the development of broader perspectives on good teaching. If this sounds open-ended, it is—by intent. There is no one way of being a good teacher; similarly, there is no one right solution, discussion, or perspective to the cases [3]. Rather the goal is to get each participant to develop his or her own successful approach

---

[1] See *Teaching and the Case Method* by L. B. Barnes, C. R. Christensen, and A. J. Hansen, Harvard Business School Press.

[2] Though most cases have only one part, a few have two parts; the second is to be distributed only when the group has analyzed the situation based on the information in the first part.

[3] This does not preclude the possibility that novice teachers will offer ill-advised perspectives; there may be no one right way to teach a first course on the integral but the Lebesgue integral is probably not the right place to begin.

to teaching. We hope that the faculty leader will facilitate a balanced and thoughtful discussion, whose direction and sophistication will depend on the participants and their backgrounds. To aid faculty in taking on such a role, which will certainly not be familiar to most, we have prepared a Teaching Guide for the facilitator (which is not intended for general distribution) for each case.

2. Preparing to Use a Case Study

Leading a case study session requires a different set of teaching skills than most mathematics teaching. The goal is to shape and guide the discussion without overpowering it or stalling it. But events happen quickly and one must be able to respond immediately. Thus teaching a case study requires preparation. Experienced case presenters in other disciplines often make many pages of careful notes in advance concerning the discussion, planning what to do if certain situations or comments arise. The notes may be organized as a tree. Though not every user of these cases will have the time and energy to prepare to this level, some advance thought is crucial.

We recommend that you first get to know the case well, reading it several times and thinking about the questions in the teaching guide. Then formulate for yourself the goals of your case session, and the three to six questions which *should be asked* in the session. Include action questions— "what would *you* do or say?"—as well as more conceptual questions. Imagine a range of participant answers or responses and plan how you will react to each [4]. You may wish to map out some signposts along the way to guide the discussion, such as some particular places in the case where an important event has occurred. Next, imagine that the discussion stalls out early, or that everyone sees things the same, in black and white. Plan what you will do or say to get it moving. For example, be prepared to ask participants "can you imagine any other way to see things?"; to challenge the group view or to take and defend an alternative point-of-view; to ask someone to model for you his or her response to the situation in the case; to specify a change in the case scenario and ask how that changes participants' reactions or what they would do then; to give an exercise to the group. Finally, plan how you will start and end the case discussion. Some possibilities are in the Teaching Guide for each case.

In conclusion, one experienced case teacher (Katherine Merseth) likened teaching a case study to launching yourself down a toboggan run. Once it begins it may be hard to keep control. But rest assured that you can guide the discussion, that your preparation will stand you in good stead—and that the ride can be exhilarating.

---

[4] Some leaders have participants respond to the case at hand by email prior to the discussion meeting. This makes it easier to anticipate participant responses and to prepare the discussion.

3. Using the Case Studies

In using the cases there are many variations: participants may read the case in advance or in the first minutes of the TA-development meeting. The group may be organized with the facilitator in front, or in a circle. The facilitator may lead the discussion and actively intervene or may play a less active role, stepping in mainly to insure that the discussion has a definite focus. Different faculty will develop different styles as case facilitators. (In our tests, graduate students believed that having a number of different faculty in a given department direct various cases would make them a more useful tool, precisely because of this variation.)

Our experiences in using the cases with varied groups give rise to the following suggestions.

1. Mathematics graduate students are not all used to talking about teaching. One can break the ice by having participants separate into small groups and carry out some specific case-related task for the first few minutes of the discussion. In the Teaching Guide for each case we suggest such a task; there are doubtless numerous effective variations. Summarizing the results for the full group starts the group discussion. (In business and law school, the first step to using a case is frequently for the facilitator to cold-call on someone to summarize the case, but our students did not like such an approach as much.) It is also possible to begin by asking students to write responses to a warm-up question to then show to the group.

2. Just as it is useful to have a structured beginning to the discussion, it is useful to have a structured conclusion, which gives students something to "take home". We have provided Wrap-Up Suggestions for each case. Having all students participate in the wrap-up was appreciated in our trials.

3. Because most of the cases have multiple themes, discussions with quite different emphases may arise out of a single case. We believe that the discussions are most useful when at least part of the discussion is focussed on specifics: how do you communicate this mathematical point, what do you do next, what specific things can you do to avoid this situation. This may require direction from the faculty member serving as the facilitator.

4. In case discussions, it is extremely useful to have some experienced voices in the room. Indeed, advanced TAs and even faculty attending the TA-development sessions can add a lot to the discussion, and can help to validate the entire exercise for the graduate student participants. So encouraging this is very appropriate. However, graduate students expressed concern that the wrong faculty member, either as leader or in the audience, could dominate the discussion or not be receptive to

alternative opinions [5]. Students were also concerned that certain faculty would judge them, perhaps negatively, by their statements about the cases. So balancing the need for experienced voices with the need for a free and open atmosphere for discussion is important.

5. It is appropriate to allow each student to participate in the discussion. In some situations, this will require intervention. For example, we have had a number of times when the quiet student offers the most useful insight, but only after being called on.

6. Just as faculty must learn new skills, graduate students (and other participants) must learn to use case studies effectively—how to respond, listen, and interact. Positive feedback to participants ("that's a nice comment") is very useful in fostering this.

7. If the blackboard is to be used, it is best to use it to record important points or to make a list of issues or possible actions, rather than to record everything said. The case leader or a selected participant may do this. In several cases, the board should be used (for example, to record the scores assigned to each answer in Making the Grade).

8. There is a learning curve to learning to lead a case discussion. It might be wise to start with a case dealing with a concrete issue, such as grading an examination, and work towards one with a more abstract theme such as motivating mathematical ideas. With each experience as a leader your skills will increase.

4. Advice and Ideas for More Experienced Case Users

As with other teaching, there are many techniques and variations which can add to your effectiveness. Here are some to keep in mind.

1. A useful technique is to pose a hypothetical variation on the case scenario and ask how that changes the analysis of the situation. When doing so, if participants are not reading the case carefully for the perspectives of the various characters, one may wish to depersonalize their responses: "you would do it that way, but what would X do?".

2. To get a participant to re-examine his or her answer or to prod the group to respond, role-play the situation for a few instants.

3. Don't be afraid of silence in a discussion—6 to 8 seconds of participants looking at their shoes may be quite healthy as they grapple with the issues.

---

[5] There is a fine line here. It is difficult to have a discussion with a leader who is perceived as saying "It's surprising you don't all agree with me" or "I guess you haven't been teaching as long as I have". By contrast, a leader may ask probing questions of participants to get them to defend their positions, or may articulately defend a position contrary to the group's, and in doing so stimulate a very successful case discussion.

4. Use your physical presence to move the discussion to the parts of the room where people are silent.

5. Don't always call on people with their hands raised (some would say don't ever call on people with their hands raised).

6. If participants are not listening to each other, ask each subsequent speaker to refer to the previous speaker by name. Alternatively, ask each speaker to mirror the speaker before, either by summarizing what was said, by repeating the last thing said, or by repeating some part of what was said. It is also possible to promote listening by breaking into small groups to discuss something, and then asking each person to summarize someone else's answer when returning to a full-group discussion.

7. If the discussion gets overheated, take a two-minute time-out from the discussion for each participant to discuss the question with his or her neighbor or to write a short analysis of the situation.

8. There are many innovative ways to use the cases. For example, some leaders will extensively mimic the situation in the case by role playing, or even set up a substantial exercise which mimics this (for example, for Order Out of Chaos, one leader designed a calculator exercise to have participants experience the students' sense of confusion). Other possibilities for the use of these cases include having students read the case in advance and do a case-related exercise or a short "reaction paper", and integrating some writing into the discussion, such as breaking in the middle to ask students to write something in response to a specific question.

9. Some case leaders record their experiences each time they use a case, for later reference. We hope to manage a website which gives such experiences. For this and other support services as they become available, please consult our website:
   `http://www.bc.edu/bc_org/avp/cas/math/publicprojectPI/`

We would welcome comments based on your experiences in using the cases. Please send them to Prof. Sol Friedberg, BCCase Project Director, Mathematics Department, Boston College, Chestnut Hill, MA 02467-3806, `friedber@bc.edu`.

Using Case Studies in a TA-Development Program

This section presents some suggestions for using case studies as part of a TA-development program and for organizing such a program. There is wide variation among institutions with regards to both the teaching duties of mathematics graduate students and the structure of their TA-development programs, so that choosing the cases which are relevant and most useful for a given program is a matter of individual institutional choice. Following the general guidelines and organizational suggestions in this section, the subsequent two sections give a brief listing of Types of Cases, sorted by teaching assistant responsibilities and by mathematics course, and a more elaborate collection of Summaries of Cases, which includes a summary of both the situation and the issues which arise in each case.

General Suggestions

1. Most TA development programs will begin with fundamental points regarding preparation, blackboard use, and professional behavior. Often the apprentice TAs will deliver practice lectures, sometimes in tandem with videotaping. At this point, it is suitable to introduce case studies.

2. Use case studies in the discussion format for about one-third of the total class time. For a course meeting two hours a week, for 30 hours in a semester, one might plan to use five cases, with about 2 hours per case. If you would like students to look at additional case studies they may be assigned as reading homework.

3. Begin with a case which has a concrete component to it, particularly if you are new to case teaching or are working with a group of graduate students who do not have previous teaching experience. For example, "Making the Grade" asks participants to grade hypothetical student work in one of 3 classes (College Algebra, Calculus I, Multivariable Calculus). "Studying the Exam" asks participants to assemble a 50-minute midterm examination from a given set of questions (once again, 3 versions). "Emily's Test" concerns possible cheating and the question is apparent: what should Emily do? Participants can make a list of options and you can discuss them in view of your institution's policies and procedures. If there is time left over you can simply ask the grad students to list ways of cheating they know of and prevention strategies for each. A somewhat more complicated possibility is "Seeking Points". This case is multifaceted, raising questions about grading, testing, and

about motivating and explaining the difference quotient, but may be an appropriate starting case if your group contains advanced graduate students.

4. Choose cases which are relevant for your TAs. As they mature, you can point out that a case may not apply directly, but the issue it raises does apply. But at first you might want cases which are immediately relevant.

5. If your TAs are not comfortable with the mathematics in a case (e.g. the definition of work, Fourier series, topics in multivariable calculus, the different parts of the Fundamental Theorem as presented in the text used by your department), you should review this briefly before using the case.

Ideas for Organization

We offer two suggestions for organizing a TA-development program. The first is based on dividing the curriculum into six topics; the second collapses these to three areas, leaving more time for focusing on basic skills.

The six topics approach: Such a course begins with blackboard work and speaking, and then discusses the topics

(1) Evaluation and Assessment

(2) Classroom Management

(3) Presenting Mathematics

(4) Technology

(5) Expectations and Attitudes

(6) Struggling Students.

Evaluation and Assessment: One goal for any teaching course must be to help graduate students prepare and grade tests, and then to deal with student reactions to the tests. This topic needs to be addressed relatively early in the seminar, because for new teachers the problems surrounding tests arise no later than the first midterm exam. The case studies "Studying the Exam," "Making the Grade," "Seeking Points," and "Emily's Test" deal with these issues. "Studying the Exam" provides a concrete set of problems from a relatively large variety of elementary courses and challenges the student to assemble a reasonable exam from the given choices. "Making the Grade" shows student work from different courses and gives students the opportunity to compare their different approaches to grading questions and assigning partial credit. "Seeking Points" treats the case of a student who is upset by the way his exam was graded and wants reconsideration, while "Emily's Test" describes a possible cheating incident. Taken together, these cases provide enough material for eight to ten hours of classroom discussion of testing issues. ("Seeking Points" could also be used in discussing the topics Presenting Mathematics and Expectations and Attitudes; "What Were They Thinking?", which concerns in part coping with poor examination results, could also be used in this section.) Additional aspects of this topic which should also be covered in a comprehensive course include making a grading curve and dealing with students who have missed an examination. Another possible topic is alternative methods of assessment.

Classroom Management: Graduate students taking a teaching seminar may be conducting discussion sections in parallel with a lecture course. You can begin this topic by challenging them on whether they should conduct their classes by managing group work by their students or by adopting a more traditional lecture/question/answer model for their classes. The two cases "Pairing Up" and "The Quicksand of Problem Four" provide the opportunity to explore these two contrasting methods. The two cases highlight some of the difficulties with both approaches. Alternatively, you might choose one of these cases and assign the other as independent reading.

Presenting Mathematics: After treating basics, assessment, and classroom management, a teaching course can turn to more mathematical concerns. Unless they are introduced to a wider range of options, apprentice teachers rely on a limited (mostly formal) repertoire to explain mathematical ideas and problems. Devoting time in a teaching seminar to expanding this repertoire can result in a significantly better ability to communicate mathematics in the classroom. For example, definitions, theorems, and problems can be explained by appeals to intuition; by geometric or physical models; by a series of related examples; and by the use of formal logical argument. Mathematical ideas may be motivated by examples, through connections to science or other parts of mathematics, or historically. The cases "Making Waves" and "Fundamental Problems" directly raise these topics, while providing motivation for learning the difference quotient also comes up in the case "Seeking Points". It is particularly useful for students to practice the techniques following the discussion: ask each to give a sample introduction to a topic they are teaching or to explain a knotty problem from their class's homework. For TAs who will teach their own course, it may be useful to complement the cases in this section by offering advice on planning an hour lecture or on planning a piece of a course (e.g. a unit on the Fundamental Theorem of Calculus).

Technology: A teaching seminar is a natural place to address the issue of the appropriate use of technology. Even if a school does not use calculators or computers as a matter of policy, they are ubiquitous. Scheduling time in class for people to learn how to use graphing calculators or symbolic algebra packages is one way to address this. Going further than technical proficiency, the case study "Order Out of Chaos" is an exploration of some of the surprising effects that graphing calculators can have in a classroom setting. You could follow up this case by asking graduate students to design their own series of calculator examples to communicate a particular mathematical idea. It would also be useful to provide information about technology support services at your institution.

Expectations and Attitudes: The cases "Study Habits," "There's Something About Ted," "What Were They Thinking?" and address some of the frustrations that teachers feel and explore some of the causes, such as unrealistic expectations and a poor perception of students' level of understanding. "Changing Sections" addresses these concerns from a slightly different point of view, dealing as it does with the issue of prerequisites and with interactions among different teachers and students in the first days of a semester.

Struggling Students: Discussing support services for students is useful. TAs should be informed of the range of institutional support available, including tutoring, time management workshops, counseling support, support for minorities and disabled students, and mental health services. The case study "Salad Days" gives a framework for discussing some of these topics, and the role of the TA in dealing with struggling students.

The three topics approach: Such a course begins with blackboard work and speaking, and then discusses the topics

(1) Evaluating Student Work

(2) TA Attitudes and Expectations

(3) Presenting Material.

Evaluating Student Work. As noted above, this topic is relevant to all apprentice teachers, and a discussion about, for example, how many points to award given student work can illustrate the ambiguity which is part of teaching. Cases here include "Making the Grade", "Studying the Exam", and "Emily's Test". Aspects of this theme are also present in: "Seeking Points", "Study Habits", and "What Were They Thinking?".

Attitudes and Expectations. The theme of how strongly the attitude of the instructor affects student performance and motivation is in almost every case. Choose one or two to illustrate this and discuss issues related to it. Particularly relevant are: "Changing Sections", "Salad Days", "Study Habits", "There's Something about Ted", and "What Were They Thinking?". Aspects of this theme are also present in: "Making Waves", "Order Out of Chaos", "Pairing Up", "The Quicksand of Problem Four", and "Seeking Points".

Presenting Material. The skill of successfully presenting mathematical material is critical. Choose a case or two which concerns this. Particularly relevant are: "Fundamental Problems", "Making Waves", "Order Out of Chaos", "Pairing Up", "The Quicksand of Problem Four", and "Seeking Points".

The case studies of this volume do not constitute a complete TA-development course by themselves. For example, as noted above you might want to discuss directly how to write a syllabus, plan an hour lecture, make a grading curve, deal with students who have missed an examination, and make use of technology support services at your institution. Two additional texts which complement the materials in this volume are Steven Krantz's *How to Teach Mathematics: a personal perspective* and Thomas Rishel's *Teaching First, A Guide for New Mathematicians*. Taking the teaching suggestions formulated in those volumes and applying them to the situations described in this one gives the prototype of the teaching experience.

Concluding Remarks

Though most TA-development programs focus on the immediate responsibilities for TAs at their institution, a graduate student who is well-prepared for postgraduate life should be ready to teach a class independently upon graduation. This readiness includes both the acquisition of basic presentation skills and also the development of more sophisticated notions of teaching, including motivating and illustrating mathematical ideas, testing fairly and thoroughly, and creating classroom conditions which maximize student achievement and learning. When integrated into a TA-development program, the case studies of this volume can play a significant role in the development of these advanced skills.

Types of Cases

The Cases presented here concern TAs (and one instructor) in a variety of settings. To find those most immediately applicable to your TAs, they are sorted below by TA responsibilities and by mathematics course involved. (Two of the Cases have multiple versions and appear on multiple lists; other Cases do not specify the exact TA responsibilities or course.) Please note that though some Cases may be more immediately relevant to your graduate students, ultimately they will need to deal with most if not all of the situations in the Cases. Thus it is not unreasonable to use a Case about a TA who teaches her own section of Calculus with a roomful of first year graduate students who will serve as recitation instructors. However, it is necessary to explain to such students the goals of the exercise and to ask them to consider the big picture as they carry out the discussion.

Cases Listed By Teaching Assistant Responsibilities

Recitation Section Instructor: Emily's Test (course unspecified); Fundamental Problems (Calculus); Pairing Up (Calculus II); The Quicksand of Problem Four (Calculus I).

Instructor for own course (possibly one with multiple sections and common supervision): Changing Sections (Calculus I); Making Waves (Calculus); Order Out of Chaos (Precalculus); Seeking Points (Calculus I); Study Habits (Calculus II); There's Something About Ted (Calculus); What Were They Thinking? (Calculus II).

Responsibilities Unspecified: Making the Grade (3 Versions: College Algebra, Calculus I, Multivariable Calculus); Salad Days (Calculus I); Studying the Exam (3 Versions: College Algebra, Calculus II, Multivariable Calculus).

Cases Listed By Mathematics Course Involved

Precalculus and below: Making the Grade (College Algebra version); Order Out of Chaos (Precalculus); Studying the Exam (College Algebra version).

Calculus: Changing Sections (Calculus I); Fundamental Problems (Calculus); Making the Grade (Calculus I and Multivariable Calculus versions); Making Waves (Calculus); Pairing Up (Calculus II); The Quicksand of Problem Four (Calculus I); Salad Days (Calculus I); Seeking Points (Calculus I);

Study Habits (Calculus II); Studying the Exam (Calculus II and Multivariable Calculus versions); There's Something About Ted (Calculus); What Were They Thinking? (Calculus II).

Course Unspecified: Emily's Test.

Two of the Cases above, Making the Grade and Studying the Exam, have a format different from the rest. These cases do not give a description of a classroom or teaching situation; instead, they are exercises (concerning grading and exam-writing, respectively), structured for group discussion. They come in College Algebra, Calculus and Multivariable Calculus versions. In some institutions they may be especially useful early on, helping to establish an atmosphere of discussing teaching issues as well as raising awareness of specific teaching issues related to these two topics.

Summaries of Cases

1. Changing Sections

Course: Calculus I.

TA Responsibility: Instructor (multiple sections with a common supervisor).

Summary: Two graduate students at a large state university, Otto and Felicia, teach parallel sections of calculus. On the first day of classes, Otto administers a test of basic skills. He advises students who do poorly on this test to drop back to to precalculus. Most of them transfer to Felicia's section instead, since they have heard that she is a good teacher and her section is offered at the same time. The last student to transfer, Gil, did poorly on the test but has other technical abilities. Felicia, in consultation with the course's faculty supervisor, must decide whether to make an exception to the rules and allow Gil into her already crowded section, or to follow Otto's suggestion and advise him to repeat precalculus.

Issues: Assessing students' backgrounds; what to do with students with weak backgrounds; remediation versus repetition of a lower-level class; the effect of students' backgrounds and their range on one's teaching; the impact of the first day on the course overall; communication between graduate students; communication between graduate students and faculty; teaching in a course with multiple instructors; section size and its implications; balancing teaching and other responsibilities.

2. Emily's Test

Course: Not specified.

TA Responsibility: Examination proctor.

Summary: A first year graduate student, Emily, is proctoring her first test. To her surprise, she observes possible evidence of a student cheating during the exam. She hesitates, unsure how best to proceed. As she is considering what her next action should be, she realizes that another student, who has subtly challenged her authority in the past, saw the whole episode, and is now watching closely to see what Emily will do next.

Issues: How to handle suspected cheating during the course of an exam; a TA's responsibilities with respect to due process in cases of suspected cheating; how a TA's interactions with a suspected cheater affect the class as a whole; when one should communicate with departmental authorities about events in one's class; how to manage a possibly confrontational situation in the classroom; preventing cheating.

3. Fundamental Problems

This case has an optional second part, which may be distributed after discussion of the first part.

Course: Calculus.

TA Responsibility: Recitation section leader.

Summary: A first-semester calculus course is about to have a midterm on the Fundamental Theorem of Calculus. In part I of the case, the TA for the recitation section, Keng, holds a review session. In it, he finds himself confronted by students with little understanding of the meaning of the Fundamental Theorem. In part II one of these students, who did well on the midterm, comes to see him. Though she can do the problems, she is concerned that she still does not entirely understand the material.

Issues: What students find confusing about the Fundamental Theorem; finding good ways to explain the Fundamental Theorem; what level of conceptual understanding of the Fundamental Theorem one should expect of students in a calculus course; what makes a good review session; how to prepare for and run a review session; how to handle student confusions when they arise in a review session.

4. Making the Grade

Course: This case comes in three different versions: College Algebra, Calculus I, and Multivariable Calculus.

TA Responsibility: Grading examinations.

Summary: In this case, participants are asked to grade sample student work from a class. They are asked to mark several problems from a given course as if they were questions on an exam, and then as if they were homework problems. During the discussion, participants will see each other's grades and discuss how they arrived at them.

Issues: Accuracy and fairness in grading; keeping point deductions proportional to the error; the role of grading in providing accurate feedback to students; the role of grading in providing accurate feedback to the instructor about the understanding of the class in general and about individual students' particular performances; the appropriateness of insisting on a particular method of solution for a problem which may be solved in more than one way.

5. Making Waves

This case has an optional second part, which may be distributed after discussion of the first part.

Course: Calculus.

TA Responsibility: Instructor.

Summary: Two graduate students, Kara and Louis, are teaching parallel sections of the same calculus course. When Louis bemoans the inclusion of Fourier series in the course syllabus, Kara points out that there are important applications of that material to waves, which she plans to show her class. Louis, unfamiliar with the applications, attends her class and is impressed. He decides to describe these applications to his own section. Unfortunately, Kara's class performs poorly on the quiz about the applications; Kara is further disheartened by a student asking her if "the physics stuff" will be on the final. Though Louis is enthusiastic, Kara questions the wisdom of her approach.

In part II, Louis is asked to take his officemate's calculus class for 2 lectures, and to cover the natural logarithm and exponential functions.

Issues: Motivating a new mathematical topic via its relation to physics or other disciplines; what content to hold students responsible for on quizzes and exams and how to communicate this to students in a way they find reasonable; balancing abstraction and theory with applications in lectures; how to manage a course with a variety of student interests and learning styles; collaboration with other lecturers in a common course; how to find enrichment material.

6. Order Out Of Chaos

Course: Precalculus.

TA Responsibility: Instructor.

Summary: Terry is alarmed to see her Precalculus section led astray by various calculator errors which distract the students from translation of functions, the topic of their exercises. She spends what seems like a long time trying to determine how the students generated various incorrect graphs. Her whole section appears mystified until, at last, a few students with good ideas speak up.

Issues: Probing student understanding and errors; developing students' ability to "reality check" their calculator output; dealing with unexpected student responses; managing class time; choosing exercises which illuminate the concept at hand; helping nurture students' mathematical intuition.

7. Pairing Up

Course: Calculus II.

TA Responsibility: Recitation section leader.

Summary: A TA running a discussion section for a course in integral calculus decides to foster her students' independent thinking skills by having them work together in pairs. She is worried that the students too often rely on her to provide them with the answers. However, the pairing exercise does not go as smoothly as she had hoped, for a variety of reasons. She is left wondering how to organize the future discussion hours.

Issues: How to best allocate discussion section time; advantages and drawbacks of group work; management of a discussion section involving group work; how to help students see errors in their own work for themselves; how to conclude a section involving group work; TA self-evaluation—how to tell if a given approach worked or not.

8. The Quicksand of Problem Four

Course: Calculus I.

TA Responsibility: Recitation section leader.

Summary: Bill Baker is the TA for a recitation section of Calculus I. According to the syllabus, the class is learning about average and instantaneous velocity. Bill plans to spend a few minutes of his section on the homework and then present some supplementary material related to instantaneous velocity. However, he discovers that most of the students in his section do not understand average velocity. Many do not seem to have put in much effort on the homework prior to section.

Issues: Figuring out what students really understand from what they say and don't say; explaining formulas and problems to students; presenting the concept of average velocity; handling a class of varied skills and preparedness; managing classroom "surprises"; balancing the students' desire to see the solutions to the homework with the teacher's desire to make students think; use of class time.

9. Salad Days

Course: Calculus I.

TA Responsibility: Not specified (either instructor or recitation section leader).

Summary: Evan is a well-intentioned first year undergraduate who is close to failing his calculus class. He is stretched too thin among many activities. He seeks relief mid-way through the semester in the form of an extension on a project, is rebuffed by his TA, and continues to do poorly. After seeking help from the TA again late in the semester in a bumbling way, Evan realizes that he must choose between withdrawing from the class, with possible financial aid consequences, or trying to squeak by with a low grade.

Issues: When and to what degree should one intervene with struggling students; what intervention is appropriate; what intervention is positive, but optional; how to be fair to all students when dealing with late assignments; providing timely and accurate feedback about student grades; how to deliver unpleasant news or have a difficult conversation with a student in a professional way.

10. Seeking Points

Course: Calculus I.

TA Responsibility: Instructor.

Summary: Daniel, an advanced graduate student, is teaching a section of Calculus I. After the first examination is handed back a surly student, Sam, comes to Daniel's office. Sam believes that he is unfairly penalized when he does not receive full credit for using the power rule, which has not yet been taught in the class, to answer an examination question which requires a derivative, rather than computing the limit of the difference quotient.

Issues: Why does one teach the difference quotient; communicating these reasons to the students; testing understanding of the difference quotient; fairness and grading; justifying partial credit to dissatisfied students; teaching Calculus to students who have already studied it in high school; the roles of practice exams and of review sessions; dealing with surly students.

11. Study Habits

Course: Calculus II.

TA Responsibility: Instructor.

Summary: Angelica, a graduate student from another country, is teaching a section of second semester Calculus. She has high expectations for the students in her class and is concerned over their poor performance to date. Angelica tries to boost her students' sagging study habits by introducing a draconian regime of quizzes, mid-semester. She is met with bemused indifference from the supervising professor, and outright indignation from her students.

Issues: Motivating students to study and do homework; the role of examinations and other methods of assessment in effective teaching; classroom management; professor/graduate instructor relations; teaching sections of a multi-section course with a set syllabus and common examinations; changing the grading contract mid-term; international graduate student expectations of U.S. undergraduates.

12. Studying the Exam

Course: This case comes in three versions: College Algebra, Calculus II, and Multivariable Calculus.

TA Responsibility: Preparing an examination.

Summary: This case asks participants to create a 50 minute exam based on a list of sample questions. Some of the questions exhibit various pitfalls a test writer might fall into, such as stating a problem unclearly, stating a problem so that poor understanding will produce a better partial answer than a slightly mistaken understanding, writing a problem which is very difficult to grade, and not considering the approaches which are possible if a calculator is allowed on the exam.

Issues: Writing a good exam: how to probe student understanding and misunderstanding, fairness, time concerns, grading concerns, use of technology.

13. There's Something About Ted

This case comes in two parts; the second should be handed out only after a discussion of the first part.

Course: Calculus.

Responsibility: Instructor (a recent Ph.D.).

Summary: A recent Ph.D. from an elite university confronts difficulties adapting to teaching at a large state school; he is confused that the lecture techniques which brought him success as a graduate student instructor are not working. In Part I, he struggles to respond to complaints from his students, getting pressure but no help from his department chair. In Part II, he hands out a survey; most of his class responds with complaints and some with hostility.

Issues: What to do when a class is not going well and when students complain; setting expectations of classroom behavior and academic performance; adjusting to a new institution's policies, standards, and student body; successfully communicating the big mathematical picture; soliciting student feedback; workload and grading issues; avoiding or changing a hostile class-teacher relationship.

14. What Were They Thinking?

Course: Calculus II.

TA Responsibility: Instructor.

Summary: Hugh Brightman, a second-year graduate student, is teaching his own Calculus II class under faculty supervision after a successful year as a TA for a recitation section in a large class taught by a professor. Although not many students have been coming to his office hours, Hugh is confident that they are well prepared for his first hour exam. He is shocked when he discovers that they don't seem to have learned even the most basic techniques and concepts.

Issues: How to gauge the real extent of student understanding; how to gauge the difficulty of an exam before one gives it; dealing with a disastrous exam; ways to raise the mathematical level of a class and make it interesting; balancing graduate mathematics studies and teaching responsibilities; the difference between the roles of recitation-section instructor and course instructor; explaining the concept of half-life.

How These Cases Were Created

Each of the cases created by this project is a fictional description of a realistic teaching situation. Unlike cases in law or business, the cases here are not direct summaries of a specific real-life experience; rather, they are fictionalized amalgams of several related experiences, written to raise some particular issue or complex of issues that is common in mathematics teaching.

The cases were created by an extensive process of writing, feedback, and rewriting. The original ideas came either from an individual development team member or from group discussion about what issues should be addressed by the body of cases as a whole. A single member of the team wrote the first draft of a case. That draft was then circulated to the development team for comment and feedback. The team met to discuss each case, then revisions were made, typically by someone other than the original author. In doing so, we strived to make the cases even-handed in their treatment of issues, as well as realistic. The revised version of each case was then distributed to a group of graduate students for written feedback. Reactions were collected and discussed in another meeting of the development team. After this discussion, someone other than the original author revised the case again, based on feedback. Some of the revisions were substantial.

The next phase of the process was to use each case with a group of graduate students and to get their reactions. Students from several Boston area universities participated in full-day meetings at Boston College. For each case, the graduate students first discussed the case in full, as if they were using it in a TA-training program at their home university, with a development team member serving as faculty facilitator. This discussion was followed by an extensive debriefing, in which all the interlocutors critically analyzed the case and discussion. The goals here were twofold: to determine whether or not the case was realistic, even-handed, and useful and if so, how it could be changed to make it most effective, and to determine what was important for a faculty member to know in order to lead a successful discussion. This oral feedback was supplemented by written evaluations of each case and of each discussion. This information was collected and analyzed in yet another development team meeting, and a subsequent round of revisions took place. At this point drafts of teaching notes for each case were prepared. These were also reviewed at later meetings and then revised.

Each case was then tested at an institution outside those of the development team members. The idea behind this stage of testing was to make sure that a case would be effective with a graduate audience which was not part of the self-selected group that participated in the initial case tests and with a faculty leader who was not part of the development team. The written feedback of participants in these discussions was once again used in deciding upon the viability of the case and upon any additional revisions.

Throughout the entire process, cases were carefully scrutinized and evaluated for their effectiveness. Several cases were scrapped entirely, while others underwent extensive revision, and were then returned to an appropriate earlier stage in the process. We hope that the cases which have emerged from this process are relevant, usable, and raise significant issues in contemporary post-secondary mathematics education.

Changing Sections
Teaching Guide

Case Synopsis

At State University, calculus is taught in sections of 30–35 students. Otto and Felicia each have responsibility for a section. On the first day of classes, Otto administers a test of skills and finds a group of weak students whom he advises to drop down to precalculus. Most of them transfer to Felicia's section instead, since they have heard that she is a good teacher and her section is offered at the same time. The last student to transfer, Gil, has a mixture of strengths and weaknesses. Felicia, in consultation with the course's faculty supervisor, must decide whether to make an exception to the rules and allow Gil into her already overcrowded section, or to follow Otto's suggestion and advise him to repeat precalculus.

Key Issues in this Case

1. What to do with students with weak backgrounds; remediation versus repetition of a lower-level class.
2. Assessing student background.
3. The effect of student background and range of student background on one's teaching.
4. The impact of the first day of teaching upon the overall course.

Secondary Issues

1. Communication between graduate student TAs.
2. Communication between TAs and faculty.
3. Section size and its implications.
4. Balancing teaching and other responsibilities.
5. Teaching in a course with multiple instructors.

Initiating the Discussion

Break students into groups. Have someone from each group defend the positions of Felicia and of Otto. Then discuss with the entire group.

Wrap-Up Suggestions

Have a student pretend they are teaching the first day of class, and address the issue of prerequisites for their class. Emphasize that the first day of class is extremely important, and that their attitude towards the class will be examined microscopically by the students.

Possible Pitfalls

1. Students may have a hard time defending a position they would rather critique. Be prepared to play devil's advocate.
2. If your institution has very different TA-TA or TA-faculty interactions than State University's, then you may need to ask the participants to find the issues in the Case which are relevant to good teaching even if some of the details do not apply directly.

Additional Discussion Questions

1. Should Felicia let Gil in or not? Put yourself in Felicia's position.
   a) First suppose you have decided to let Gil into your class. What would you do to help him catch up?
   b) Now suppose you have decided to tell Gil he can't add the class. Give your explanation of your decision.
2. Should Otto have advised Gil to drop the class? Take the role of Otto and give his explanation to Gil of why he should drop down to precalculus.
3. What are Gil's true abilities? What would be best for him, to go back and 'get it right' or to muddle along?
4. Otto and Felicia are both successful teachers. Whose students benefit more in the long run, Felicia's or Otto's? [Participants may think that Felicia's popularity is merely the result of her willingness to accommodate weak students, or that Otto's success on the common final is merely the result of weeding out the weaker students at the beginning. Have them look for positive aspects of each approach as well as negative.]
5. What do you think of Walter's approach to the dilemma? Was he right to insist that Felicia cap her section size?
6. Should students have been allowed to change sections? Is the system that allows this fair to TAs?
7. What are appropriate goals the first day of classes? How does one accomplish them? (Note: beginning teachers may not appreciate how much of their students' attitudes towards the class are determined by the first day. Different graduate students will have different goals for this first day, from establishing their own authority and inculcating a view of the class as a place for work and learning to establishing a sense that students in this class can succeed in spite of past difficulties and a sense of availability and collegiality. Communication of prerequisites and of course management details are other clear first-day goals.)
8. Whose responsibility is it to decide what to do with under-prepared students? The university's? The department's? The course supervisor's? The TA's?

Emily's Test
Teaching Guide

Case Synopsis

Emily, a first year graduate teaching assistant, is proctoring her first test. In the course of doing so she observes possible evidence of a student cheating. She hesitates, unsure how best to proceed. As she deliberates, she realizes that a student who has been a source of frustration to her is watching her to see what steps she will take.

Key Issues in this Case

1. What is the best way to handle cheating when it has been observed or when it is suspected?
2. What are a TA's responsibilities with regard to due process in cases of suspected cheating?
3. How does a TA's actions towards one student who may be cheating affect the rest of the class?

Secondary Issues

1. What strategies can be used to prevent cheating from taking place?
2. When does one need to communicate with departmental authorities about events in one's class? How?
3. How can one maintain control in a confrontational classroom situation?
4. What are some strategies for dealing with students with a negative attitude?

Initiating the Discussion

Separate the participants into groups of 2 or 3 and ask each group to consider Emily's options and decide what Emily should do. Have each group list the actions Emily could take, and decide what they think is best. As the groups report on their decisions make a master list of the different possibilities and choices.

Wrap-Up Suggestions

Provide written documentation concerning your institution's policies and procedures for cases of suspected cheating.

Possible Pitfalls

1. The group may not adequately consider possible explanations of Liam's behavior which do not involve cheating. In this case, the discussion leader might wish to actively solicit such explanations or to provide them. In considering possible responses to the situation (for example, taking Liam's paper away), it might be helpful to note actions that Liam, his family, or

even his lawyer might take in response (for example, a call to the department chair).

2. The group may not be aware of institutional policies concerning cheating and due process. We recommend that the facilitator provide this information to the group, either at some point during the case or at the conclusion of the discussion.

3. If the group reaches an early consensus on this case, one may want to direct the discussion into a discussion of the different kinds of cheating and the ways that they may be prevented. Have the group make a list of methods of cheating and corresponding methods of prevention. In field testing, such a discussion was considered especially valuable by the participants.

Additional Discussion Questions

1. What innocuous explanations are there for Liam's behavior?

2. Were Emily's instructions concerning taking the examination sufficient? Could they have been improved? Is it necessary to have written class policies concerning examination procedures? Concerning cheating?

3. What should Emily tell her supervising professor about the situation? How important is this? Will she be supported by her professor? By her department chair?

4. What are institutional policies concerning cheating? How much variation is there among institutions? Why?

5. Do any members of the group have any experiences with cheating at this institution? At other institutions?

6. What are other ways and situations in which cheating can occur? Discuss possible ways of avoiding each type of cheating.

7. For graduate students who will teach their own courses: under what circumstances is it appropriate to give a penalty in a grade for suspected cheating? What penalties are deemed appropriate for various offenses? How is one expected to notify the student? Who else must be notified? What procedures must be followed in order to preserve the student's rights and for a penalty to be upheld on appeal?

Fundamental Problems

Part II

At Keng's next section, he went over the exam, which had just been handed back. Since there wasn't any new homework, Keng asked if there were any questions and when there were none, he dismissed the class 10 minutes early. As the students were leaving and Keng was gathering up his notes, Lindsey came up to him, somewhat hesitantly. She had gotten an 89 on the exam, an A-.

"You know, Mr. Keng," she said, "I know how to do the homework problems in Chapter 5. But I'm still having a hard time with the difference between definite and indefinite integrals, and I don't really understand why the Fundamental Theorem is so important. Could you please explain it all to me sometime?"

Fundamental Problems
Teaching Guide

This case is designed to be used in two parts. The second part of the case should be distributed only after discussion of the first part; it is not included in the graduate student edition.

Since not all participants may be familiar with the way that the Fundamental Theorem is presented in the Calculus texts your program uses, it may be useful to ask them to read these sections before doing this Case.

Case Synopsis

A class in Calculus is about to have a midterm on the Fundamental Theorem of Calculus. In part I Keng, the recitation section TA, holds a review session. In it he finds himself confronted by students with little understanding of the meaning of the Fundamental Theorem. In part II one of these students, Lindsey, does well on the subsequent midterm but believes that she still lacks full understanding of the material.

Key Issues in this Case

1. What are the sources of confusion for students in the Fundamental Theorem?
2. How does one explain the Fundamental Theorem, including the relation between different parts or versions of the Fundamental Theorem?
3. In a Calculus course, what level of conceptual understanding can or should be expected of the Fundamental Theorem?

Secondary Issues

1. How does one prepare for a review session? How does one run such a session?
2. What strategies can one use in order to handle a range of questions and of understanding in a review session?
3. How can one use questions to figure out why a student is confused?
4. How should one deal with a topic in a review session when students are confused about it?

Initiating the Discussion

Divide participants into groups of 2 or 3. Ask each group to list the (mathematical) reasons that Lindsey may be confused. Ask participants to describe:

(a) What Lindsey heard and how that could have led to her confusion.

(b) What questions they would ask Lindsey to determine if their picture of her understanding is correct.

(c) What they would say to Lindsey (and others at the review session), assuming that their views of her understanding are correct.

Then ask similar questions concerning the understanding of the student in the back of the room.

Wrap Up Suggestions

Review responses to teaching the Fundamental Theorem, focusing on avoiding student misunderstandings and on developing undergraduates' knowledge of and appreciation of the theorem.

Possible Pitfalls

1. Teaching the Fundamental Theorem well is challenging for even the best faculty. Graduate students may need to be told this.
2. The student misunderstandings described here may not be familiar to inexperienced teachers. Experienced voices may need to confirm that they are common.
3. The main issues may arise in the discussion of part I. In that case, part II may be dealt with briefly or omitted.

Additional Discussion Questions

1. *Developing a Picture of Lindsey's Understanding.* Using ideas from the groups, piece together a description of Lindsey's understanding of the Fundamental Theorem. Draw out that she has apparently heard a definition of an indefinite integral in lecture (in terms of antiderivatives) and, because the symbol for indefinite and definite integrals are similar, has confused the two.

   Avoid settling for the explanation that the Professor either made a mistake or was unclear in the lecture, as the type of misunderstanding described frequently occurs even after the best lectures.

2. *Checking the Picture of Lindsey's Understanding.* Get the TAs to think of questions to ask Lindsey to validate their hypothesis about her understanding. If the groups' views are divergent, ask someone from one group to respond as Lindsey would have done to questions from another group. Draw out the importance of questioning to check students' understanding, and discuss the possible consequences of not doing so.

3. *Developing and Checking a Picture of the Other Questioner's Understanding.* Similarly, what are the sources of confusion for the student in the back of the room? What might he know? What might he be confused about? Is his question likely to reflect confusion on the part of many students? How could this be checked?

4. *What to do Next?* Collect ideas of how Keng should proceed. Different hypotheses of students' views may lead to different approaches. Also, collect ideas of what Keng might have done differently in preparing for and organizing the review session.

5. *The Midterm Exam.* What do you think the exam that the students took looked like? What suggestions do you have for the exam? Was Lindsey's exam grade reasonable?

6. *Responding to Lindsey after the Exam.* What should Keng say to Lindsey? Does her concern reflect a problem with the course?

7. *Developing Student Appreciation for the Fundamental Theorem.* Have participants give explanations as to why the Fundamental Theorem is indeed fundamental, as they would to their own classes.

Making the Grade
Teaching Guide

Case Synopsis

This Case provides sample student work from a class. The participants are asked to grade the work, first on a 10-point scale, then on a 3-point scale. There are 3 versions of the Case, giving work from a class on College Algebra (a review of algebra and its applications in a basic mathematics class), Calculus I, and Multivariable Calculus.

Key Issues in this Case

1. Accuracy and fairness in grading.
2. Keeping point deductions appropriate to the error.
3. The role of grading in providing accurate feedback to students.
4. The role of grading in providing the instructor with feedback on student understanding in general and on each individual student's performance in particular.

Secondary Issues

1. Appropriateness of requiring a specific method of solution for a problem which may be solved in more than one way.
2. Policies and procedures at the institution for cases of suspected cheating (this issue is addressed directly in the Case *Emily's Test*).

Initiating the Discussion

Allow the participants 15 minutes or so to grade the offered problems, noting that participants need not make detailed comments on the student work. When the participants are finished grading, make two large tables on the blackboard for question 1. The tables should have 3 columns, one for each of students A, B, and C. One of the tables will be for the scores out of 10 points, one for the scores out of 3 points. Write the numerical grade assigned by each participant to each of the three students in the appropriate columns. (This Case works best if participants see the group's entire set of grades written on the blackboard prior to the discussion.) Then have the students discuss how they assigned the grades for each student, in both the 10-point and the 3-point scenarios. For any scores that don't seem appropriate or which don't fit the popular pattern, ask the individual participants to describe their reasoning.

Repeat for question 2.

Wrap-Up Suggestions

1. Discuss different grading techniques which help ensure accuracy, consistency, and fairness, such as

   - writing an accurate scoring rubric for each problem on a quiz or exam and keeping detailed notes of deductions made for various errors;
   - grading problem 1 for each student in the class before going on to grade problem 2 (on a test or homework), as opposed to grading each student's entire paper in sequence;
   - trying a "positive" approach to grading of assigning points to a student who correctly completes each step in a given problem instead of a "negative" approach of subtracting points for various types of errors.

2. Have the participants list ways in which they can increase their accuracy and fairness in grading a large class.
3. Have participants list the purposes grades serve from the student's perspective and from the faculty perspective. Have the grades they assigned served these purposes?

Possible Pitfalls

1. Some participants may have trouble separating the 3-point scale from the 10-point scale in their minds. Have participants discuss whether these scales need to be proportionate or if it's OK to use different sorts of scoring rubrics for different purposes.
2. The discussion may be polarized if one student is far from the others. Remind the participants that in grading one should strive for consistency and fairness within one's own classes but that experienced faculty disagree on how to handle many of these situations.
3. Some participants may give all student work essentially the same grade (either all high grades, or all half-credit). This provides an opportunity to discuss the feedback a student gets from the grades received. Ask the following questions:

   - What topic or technique is being tested in the problem?
   - Which students understand this topic? Which do not?
   - What message does a grade of 5/10 (resp. 7/10 or 8/10) send to a student? What message do we wish to send to this student? What grade would send that message?

4. If the entire group grades the problems in the same way, you may wish to play a contrary role and defend a different set of grades than the group's. Think about this in advance of using the Case.

Additional Discussion Questions

1. Some participants may use fractions of a point in their grading. Is this wise? Appropriate? Time-effective?

2. What sorts of standards are appropriate for quizzes or exams? What sorts of standards are appropriate for homework? How can we develop rubrics appropriate to these standards?

3. In grading papers, graduate students may notice correct answers which are unjustified by (or even contradictory to) any supporting calculations or two papers which are suspiciously similar. Are participants aware of the institution's policies and procedures for handling suspected cases of academic dishonesty or academic misconduct? If not, you may wish to either present this information or assign participants to research this for later discussion. Under what circumstances is it appropriate to give a penalty in a grade for suspected cheating? What penalties are deemed appropriate for various offenses? How is one expected to notify the student? Who else must be notified? What procedures must be followed in order to preserve the student's rights and for a penalty to be upheld on appeal? (These issues are dealt with directly in the Case *Emily's Test*.)

Specific Remarks: College Algebra Version

1. Some participants may assign the same grade to all three students A, B, and C in problem 1 or may assign a rather high grade (7/10 or 8/10) to student B on problem 1. As noted above, one should ask which students understand this topic, and which do not.

2. Some participants may think that the algebraic error made by student C in problem 1 should receive a large point deduction. Others may believe that student C deserves no point deduction on problem 1, since the problem was correctly solved a few steps before the error occurred. To help participants come to an accurate perspective of your institution's student body and the appropriate expectation instructors should have at each level of instruction, try asking the following questions from the faculty perspective:

   - Which student knows the most about the topic being tested? Which knows the least?
   - What are the goals for the course? How close are the students to meeting these goals?
   - How do the assigned scores rank the students? How do we wish them to be ranked? If we were reviewing these scores a few weeks later, what information would they give us? What information would we like them to give us?

3. In problem 2, participants may disagree on how to grade student B, whose answer could be easily checked to be wrong.

4. In problem 2, participants may also disagree on how to grade student C, who doesn't use the algebraic method covered in class. Some participants may not be aware of the current emphasis on student dis-

covery in most American high school curricula. You may wish to refer participants to the National Council of Teachers of Mathematics (NCTM) web site http://www.nctm.org/standards/ for a discussion of the NCTM Standards, or have younger participants discuss their own experiences with discovery-based approaches to problem-solving, particularly in lower-level college classes or in high school. Discuss how one might word homework, quiz, or test questions in order to require a particular method of solution, and the advisability of doing so in this instance. (You might also ask any participants who are experienced teachers to describe how often they see student solutions which exhibit creativity, or even common sense.)

Specific Remarks: Calculus I Version

1. It is frequently helpful to begin assigning grades by asking what topics the given problem tests. For example, in problem 1, before reading the student work the participants might have mentioned only the chain rule, but after reading the student work, it is apparent that the problem also tests the student's knowledge of trigonometric functions and their derivatives. Partial credit should reflect this. It may be useful to mention to participants without grading experience that it is almost always helpful to read several papers before making a scoring rubric.

2. Participants may disagree widely on how to score some of the student work. You might wish to ask participants to describe how they give partial credit; in particular do they start at 10 and subtract points for various sorts of errors, or do they start at 0 and add points for steps in the solution which are completed correctly? The difference between these two views may be quite pronounced in some participants' grades for some of the students, in particular for students C in problem 1 and B in problem 2. Which approach is preferable? Which yields more appropriate grades? In discussing this, one might wish to ask

   - How much do the students know about the topic being tested?
   - Do the assigned scores reflect the amount of the correct solution which is present on the students' papers? Should they?
   - How do the assigned scores rank the students? How do we wish them to be ranked? If we were reviewing these scores a few weeks later, what information would they give us? What information would we like them to give us?

3. With regard to the work of student D on problem 1, some participants might feel that their grade for this student would depend on the rest of the student's exam or on how good a student this person was known to be, reasoning that they would be more likely to consider this a careless error for a student known to be a good student. It may be useful to ask the participants about their view of the role of impartiality in exam

grading. Should grades for identical work depend on the identity of the author? Should the grades we assign reflect primarily the quality of a student's written work, or should they also reflect our judgment of the student's overall ability or potential?

4. The work of student B in problem 2 raises the question of how one should handle errors, even minor errors, which cause major changes in the difficulty of a problem. This is a problem which arises with surprising frequency. The questions in item 2 above may be of use in coming to a decision.

5. With regard to the work of student C on problem 2, participants may disagree on how much credit to give for the unexpected use of an alternative solution method, L'Hôpital's Rule, before it is taught. If so, you may wish to ask participants about the appropriateness of grading for a specific solution method if none was requested, and about how one may write exam questions to require a specific solution method. The issue of knowledge from prior classes also arises here: what do we wish students to use and what do we wish them not to use?

Specific Remarks: Multivariable Calculus Version

1. At some institutions beginning graduate students may have better backgrounds in single variable calculus than in multivariable calculus, and some may not feel confident in grading the given problems. If so, it might be helpful either to review the problems briefly at the beginning of the discussion or to ask in advance that students review these topics on their own.

2. Some participants may grade students A and B essentially the same way on problem 1, since at first glance both have omitted just one portion of the formula for the tangent plane. One should ask if both students understand the topic being tested. How do we handle seemingly minor errors that suggest major misunderstandings?

3. Some participants may give student C full credit on problem 1 for the correct answer, even though there is an error in the previous step. One could then ask what message we wish to send this student about his/her mathematical knowledge and which grade will send that message. Does this student understand that tangent planes are linear, or what linearity means?

4. In grading problem 2, it is important to discuss how to take into account the consistency between the level curves and the three-dimensional graph, and the consistency of the three-dimensional graph with the original function (which is, after all, quadratic and non-negative). Try asking participants how their grade on problem 2 for student B would change if the three-dimensional graph instead were a "bowl" which agreed with the given elliptical level curves.

5. In problem 2, some participants may point out that they cannot tell where student B got his/her three-dimensional graph, since it does not agree with the given level curves, and some may suggest that it might have come from the paper of student B's neighbor. See the comments on academic dishonesty (additional discussion question 3 above).

6. Participants may disagree, particularly with respect to the work of student A on problem 2, about the weight which should be given to errors which are purely algebraic in a multivariable calculus class. That work also raises a tricky question which arises surprisingly often: how should one handle errors, even minor errors, which lead to major changes in the problem? Some useful questions to keep in mind are:

   - How much does the student know about the topic being tested?
   - Do the assigned scores reflect the amount of the correct solution which is present on the students' papers? Should they?
   - How do the assigned scores rank the students? How do we wish them to be ranked? If we were reviewing these scores a few weeks later, what information would they give us? What information would we like them to give us?

You may also wish to point out that the graphing discussion in a multivariable calculus class often includes a "catalog" of different functions and their graphs which may give some emphasis to saddle surfaces, so these are likely to be on students' minds and the first thing which occurs to a nervous—or an underprepared—student.

Making Waves

Part II

The following Tuesday Louis found a note from his officemate in his mailbox:

Louis,

My wife's father died unexpectedly, and we are leaving right away for the funeral. Would you please cover my Intro to Calculus lectures tomorrow and Friday, at 9 a.m. in room 371B? The topics are the logarithm and the exponential functions, including the limit definition of the number e. I really appreciate it, and will take your classes for you to make it up when I return.

Sam

Louis picked up his officemate's course textbook and began to think. He wondered what Kara would do.

Making Waves
Teaching Guide

This case has an optional second part, which may be distributed after discussion of the first part. The second part is not included in the graduate student edition.

Case Synopsis

Kara and Louis are teaching sections of the same calculus course. Kara mentions using applications to motivate the introduction of Fourier Series. Louis, unfamiliar with the applications, attends her class and is impressed. He decides to follow a similar approach in his own section. Then Kara's class performs poorly on a quiz which covered the application and further one of her students has discouraged her by asking if the "physics stuff" is going to be on the final exam. Kara questions the wisdom of her approach. In part II, Louis is asked to take his officemate's calculus class for 2 lectures, and to cover the logarithm and the exponential functions.

Key Issues in this Case

1. Introducing a new mathematical topic with motivation, including relations to physics or other disciplines.
2. Is this going to be on the test? What content to hold students responsible for in quizzes and examinations, and how to communicate and justify this to the students.
3. Balancing the teaching of theory and abstract content with the teaching of applications.

Secondary Issues

1. Managing a course with a range of student interests vis-a-vis the course.
2. Collaboration with others teaching a common course.
3. Identifying resources which allow one to enrich the textbook's material.
4. Alternative means of assessment, particularly in relation to non-core material which is nontheless important.

Initiating the Discussion

Separate the participants into groups of 2 or 3 and ask each group to identify the main issues the case brings to mind. Have each group report their choices. Develop the list on the board or overhead projector from the group responses. Allow for additional items to be suggested by the whole group. Have the group prioritize the list.

Wrap-Up Suggestions

Save the last ten minutes of the training session for one of the participants to model an introduction to a standard calculus topic, with the others to provide feedback.

Possible Pitfalls

1. One should be prepared for the graduate student who thinks it is impossible to get students to appreciate such extra material. Also, some students may come from a different culture where students are less likely to ask "what is it good for?", and may not appreciate the importance of this question in most American classrooms.

2. Not all institutions mention Fourier series in a lower division course. If yours does not, you may need to remind participants that the Case raises issues which are relevant to good teaching even if the details do not apply directly.

3. If the discussion centers on motivation, everyone will likely agree that some motivation is good but that it must be balanced with other aspects of the course. The discussion leader then needs to guide the discussion to specifics, asking participants to motivate particular concepts. One way to do this is to move to Part II, which poses such a challenge. An alternative is to ask to the group to list approaches to motivating course topics (see question 2 below), and then ask the participants to motivate important concepts from the classes they themselves are working with (question 3 below). Have students present this as they would to their own classes.

Additional Discussion Questions

1. What motivational "extras" are crucial for a calculus course? A precalculus course? A differential equations course? A linear algebra course?

2. How does one motivate a topic, definition, or theorem? (Answers include bringing in connections to other disciplines such as physics, computer science, and economics, mentioning connections to other mathematical topics or areas which students may learn later, designing a careful set of examples leading up to a definition or theorem, and explaining what the topic in question is used for (e.g. polar coordinates are used to coordinatize situations in which there are circular symmetries, such as objects spinning or rotating).)

3. Choose a topic in a course you are TAing for. Motivate the topic. (For TAs doing a recitation section of a course, imagine a student coming to recitation section with the question "what is this stuff good for?".)

4. Is it appropriate to test introductory material relating the math to other disciplines on a math test? Why or why not? If it is appropriate, how

would you do so? (One may make this question specific to the course the student is TAing.)

5. Are there other ways one could strengthen students' motivation to learn the material presented?

6. Think of the strongest students in a class you are teaching or have taught. What can you do to increase their interest in mathematics? How can you successfully challenge them?

Order Out of Chaos
Teaching Guide

Case Synopsis

Terry is teaching her own section of a pre-Calculus class. During a previous class meeting, she introduced the idea of horizontal translations of functions beginning with parabolas, and assigned a set of functions to be graphed for homework. To begin today's class meeting, she asks for volunteers to put sketches of some of their graphs on the board. Unexpectedly, some of the sketches are incorrect, and the ensuing discussion reveals that more than a few students are unclear as to the relationship between the graph of a function, $f(x)$, and the graph of its translate, $f(x - a)$.

Key Issues in this Case

1. Probing student understanding and errors.
2. Developing students' ability to "reality check" their calculator output.
3. Time and classroom discussion management.

Secondary Issues

1. Additional approaches to developing student understanding of translations and more generally, transformations.
2. Relation of this Case to teaching Calculus.

Initiating the Discussion

Divide the participants into groups of 2 or 3 and ask each group to develop a plausible explanation of how each student in the Case arrived at the table or picture or statement that he/she made. This can lead naturally to a discussion about how to find out what students in the class are thinking, and then how to correct errors in thinking. Then have groups identify "teachable" moments in the Case and develop a strategy to capitalize on each.

Wrap-Up Suggestions

The Case ends on a somewhat hopeful note with Marilyn's comment indicating her attempt to troubleshoot the graphs put up by her classmates and to reconcile them with what she believes should be the characteristics of the graphs of the functions. Case leaders could ask participants (possibly in groups) to develop a strategy for wrapping up the class discussion by playing off of Marilyn's comment. (The uncertain tone of Alex's comment could also be used.) Additional possible wrap-up questions are given below.

Possible Pitfalls

1. Participants in the Case can sometimes become wrapped up in only the first few student voices, without thinking much about other student voices in the Case. Direct questioning of participants, e.g. "What do you (name a participant) think of 'so and so's' comment or understanding?", can help move the discussion along.

2. Discussion has the potential to "hang up" on an overanalysis of the graphs themselves and/or the Ralph's table. Case leaders should be prepared to intervene.

3. The usage of calculators in teaching is a hot issue and the discussion may polarize, with all participants taking one side or the other. The discussion leader may need to encourage contrary views and actively bring out the teaching issues raised by the Case.

These pitfalls make this Case more challenging to use than some of the others. It is recommended for more experienced case leaders.

Additional Discussion Questions

I. Responding to Specific Moments in this Case

1. What is your assessment of what Terry did after Ralph and Natasha first finished putting their work on the board? What else could Terry have done either in addition to or instead of what she did? What would you do?

2. How might Terry deal with Ralph's table and help him and others see the probable source of the error and what is going on?

3. How might Terry treat Ralph's response about "flipping" his graph, and Jason's follow-up comments?

4. What transformation does Ralph's picture depict? How might Terry seize on this moment to engage the class in a discussion of this transformation?

5. The events present the potential for Terry to engage the class in a discussion about vertical translations, although she may have wanted to begin with horizontal translations. Where does this occur, and how would you have handled the situation?

6. What are some other "teachable" moments in this exchange, and when and how would you intervene?

II. Classroom Management

1. How long should a discussion go on? When and how do you intervene? Does it matter how much time is spent on a particular problem? How do you handle student errors in you section or classroom? Is control of the class an issue?

2. What strategies have you used (can you use) to elicit a sense of what other students in the class are thinking? Do you think the level of understanding of the students who spoke up are typical of the students in this class?

III. Communication of the Mathematics

1. What set of experiments/activities would you design to explore transformations using a graphing tool, and also to highlight a "responsible" use of that tool?
2. What strategies can you use to induce students to think about their calculator output rather than just blindly believe it? What "calculator counterexamples" have you used, or could you develop?
3. What other approaches to exploring transformations could you use? What are some of the relative merits of these approaches?
4. What do you think the point of the homework assignments was? Do you think the students understood this point? What strategies could you follow to (i) articulate the goal of an assignment, and (ii) ascertain whether the students understood it or not?

IV. Wrap-Up Questions

1. What could Terry have done before class to avoid the situation altogether?
2. If you were observing Terry's class, what would you discuss with her?
3. Has the situation depicted in this Case ever presented itself to you? If so, how have you dealt with it?
4. How would you continue the class from the point at the end of the Case?
5. If there were only 5 minutes left in the class period, how would you sum up the class? What activities and/or questions could you assign that students could think about and work on to enable you to pick up the discussion at the next class meeting?

Exercise for: Order Out of Chaos

Terry starts the class by asking students to put some graphing problems from the previous day on the board. In this exercise we will consider some different sorts of graphing problems she might have assigned.

(a) Design a sequence of graphing problems Terry could have assigned that might have skirted the confusion some of her students had. Bear in mind her interest in showing the geometric meaning behind the algebra.

(b) Instead of trying to design problems that avoid the traps her students fell into, Terry might have wanted to give some that revealed the mistakes they were making. Choose either Ralph's or Jason's erroneous graph and think of some follow-up graphing problems you could have given each of them to help them see their mistake.

(c) Marilyn had a key insight that enabled her to avoid the problems Ralph and Jason had. Design some graphing problems that would lead students towards that insight.

Pairing Up
Teaching Guide

Case Synopsis

Shalini is a TA running discussion sections for a course in integral calculus. Concerned about her students' lack of independent thinking and reliance on her to provide all the answers, she decides to experiment with having her students work on problems in pairs. The experience doesn't go as smoothly as she had hoped, and she must decide how to spend her coming discussion sections.

Key Issues in this Case

1. Discussion section management; how is section time best spent?
2. Helping students solve problems and detect errors while working in groups.
3. Managing a discussion section involving group work.

Secondary Issues

1. Benefits and drawbacks of various team pairing arrangements.
2. TA/student interactions.
3. How to best conclude a section.
4. TA self-evaluation; how to determine what is helping the students.

Initiating the Discussion

Begin by asking two selected students for a summary of the main features of the Case. This should feed into a discussion about how to view/handle the various interactions in the case. It is helpful to ask students what they would do if they were Shalini in the situations she encounters.

Alternatively, pair participants and have each pair examine the interactions between Shalini and one pair of students in the Case—what Shalini did well and what not well, what she might have done differently, etc.—for about 10 minutes. Then compare observations.

Wrap-Up Suggestions

Imagine that Shalini had time for 5-10 minutes of closure at the end of her section. Ask students what they would have her do and/or say. Ask them what homework she should assign for the next section.

Possible Pitfalls

Not all institutions cover work in their calculus classes. (Even some mathematics graduate students might benefit from a quick review.) You may need to remind participants that even if the particular example does not apply, there is still much to be learned from the scenario described here.

Additional Discussion Questions

I. Communication of the Mathematics

1. What could Shalini say to Trina and Lucas to help them realize that their answer is wrong? Could she ask a leading question about the particular problem in a better way? Should she be explicit?

2. Is there a way Shalini might draw an analogy between this problem and a kind of problem the students already know how to do? If there had been a few more minutes, and Cassandra had seen that different slices have different radii, but not how to apply this knowledge, what might Shalini have said to her?

3. Suppose that next section, Trina and Lucas realize that different slices need different amounts of work to move them, but do not know how to translate this information to an integral. What should Shalini do?

4. If Shalini had spent the last 10 minutes explaining how to do the problem and then assigned each group a similar problem with a different shape (hemisphere, etc.) as homework, what responses would you expect her to get next time?

5. What should Shalini do in her next class?

II. Teaching and Classroom Issues

1. Is discussion section time well spent in having students work in pairs like in Shalini's section?

2. What are the benefits and drawbacks of various pairing arrangements— e.g., having students of the same skill level work together, letting the students pick for themselves, or having the TA randomly assign partners?

3. Shalini often leaves her students in the middle of a discussion. What might make this appropriate or inappropriate?

4. How would you have responded to each of the three student pairs?

5. There is no closure at the end of the class. Must every class end with a summary or presentation?

6. Shalini sees only the negatives in her performance. Have there been any positive results?

7. Should Shalini offer an alternative problem to better engage the more mathematically sophisticated students? If so, what? If not, how can she keep their interest, or even better, stimulate it?

8. Should Shalini continue with work in pairs in the next class? If so, should she keep the same pairs? If not, what should she do and how should she explain it to the class?

9. Should Shalini hand out a worked solution to the problem as the students leave her class? At the next class? What are the pros and cons of doing so?

As a possible follow up to the discussion of this Case, one could have students read some of the literature concerning cognitive models for learning mathematics or concerning mathematics pedagogy and then revisit the Case.

The Quicksand of Problem Four
Teaching Guide

Due to the length of this Case, it is recommended that students be asked to read the Case beforehand, or else that an hour and a half be allocated for the presentation.

Case Synopsis

Bill Baker is the TA for a recitation section of Calculus I. According to the syllabus, the class is learning about average and instantaneous velocity. Bill plans to spend a few minutes of his section on the homework and then present some supplementary material related to instantaneous velocity. However, he discovers that most of the students in his section do not understand average velocity. Many do not seem to have put in much effort on the homework prior to section.

Key Issues in this Case

1. Correctly interpreting how well students understand material based on what they say or don't say in class.
2. Classroom management: managing a class discussion so that a variety of students participate and benefit. Anticipating the range of student responses to questions and preparing accordingly. Balancing the students' desire to see the solutions to problems with the teacher's desire to make the students think.
3. How should the concept of average velocity be presented? Explaining concepts, formulas, and problems.

Secondary Issues

1. Handling students of different ability levels in a discussion.
2. Managing classroom "surprises."
3. Teacher's emotional response to classroom developments.

Initiating the Discussion

Break the students into small groups and ask them to review the case. Ask them to identify the key aspects and difficulties in this first half of the discussion section. After 5-10 minutes, have each group summarize their findings while the facilitator writes them on the board.

In a second phase, ask the groups what Bill should do next, and/or what he should have done differently.

For alternative approaches and additional questions, see the more detailed notes on the following pages.

Wrap-Up Suggestions

Ask each participant for a short answer to one of the questions given in the section *Additional Discussion Questions* below (see especially part III), or ask each participant to state the most useful suggestion which arose in the discussion.

Possible Pitfalls

1. The case covers a lot of ground, and it is easy to wander in generalities.

2. Participants may focus on the management issues and avoid the important question of how to explain the mathematics.

3. Participants like to beat up on Bill, blaming him for the problems he faces. Be sure to draw out positive comments and constructive criticism.

Alternative Ways to Organize the Discussion

1. Have participants select a character in the case and write a student evaluation for Bill from that character's point of view.

2. Have participants play the role of a Professor who has just observed the section described in the case and who is trying to help Bill improve his teaching.

3. Have students read some of the literature concerning cognitive models for learning mathematics, and analyze the situation using them.

Additional Discussion Questions

The following questions arise out of the case material and can be used to stimulate a more focused discussion of various aspects of the case.

I. Communication of the Mathematics

1. *Why is average velocity difficult for Bill's students to understand?* Have participants play the role of Bill explaining average velocity to individual students from the case.

2. *How do Bill and his students use and understand the formulas involved?*

   a. Explain the symbols in the formula for v_{avg}. Explain the role of coordinatization in the solution to the homework. What would you say to a student who was confused about slope in s, t rather than x, y coordinates?

   b. How do you want students to use and understand such formulas? How can you help them do this? How important is this?

    c. More generally, how could Bill get his students to think about and understand the material? What would you do?

3. *What assumptions does Bill make about his students and what's going on in their minds?*

    a. How reasonable are these assumptions?

    b. What are the consequences of his assumptions?

    c. How could Bill test his assumption that students learned about velocity in class yesterday, and find out whether or not this material made sense to the students? (For example, Bill might ask the students directly whether or not they were satisfied by the professor's explanations of average and of instantaneous velocity.)

4. *Have you ever had a similar problem?* How do you engage students with the mathematical ideas you teach in ways that are meaningful to them?

5. *What is the root cause of the apparent disconnect between Bill and the class?* This is the fourth discussion section of the semester, and in Bill's mind, things to this point had been progressing "reasonably well". (Some possible answers: The material might be becoming more difficult, the professor might have failed to present the material on velocity coherently, or perhaps the students have been struggling more in past weeks than Bill has observed.)

6. *What is Bill's attitude towards the mathematics he is covering?* (Note page 3 bottom, "this was pretty basic stuff. He wanted to get through this so he could talk about the interesting things".)

    a. How does Bill's attitude compare with the students' attitudes?

    b. If they are different, should Bill address this directly? If so, how? (As one idea, Bill might present a brief historical context to explain the importance (and difficulty!) of the material.)

II. Discussion Session Management

1. *Is Bill's plan to conduct his discussion session interactively a good one?.*

    a. To what extent is he successful? Has Bill created an atmosphere which encourages interaction?

    b. What warning signs are there that the students are struggling with the material?

    c. In what ways could Bill improve in this area?

Remarks:

  (i) Note Bill's success in getting students to give answers and what he has uncovered about their lack of understanding of an important concept. Participants might take this for granted, but it is actually quite impressive!

(ii) Note students shifting in their seats and reluctance to raise hands on page 1. This raises the question of how participants react to and interpret silence in response to their questions.

(iii) Note calling on Kathy without warning (towards the end of the Case); is this a good idea? Why can't Kathy answer the question Bill asks her? Will she be likely to volunteer answers in the future? What about cold-calling on students in general; why or why not? (This can lead to a discussion about finding out what all the students know, efforts to involve all students, and balancing the comfort of the student with the exposure necessary for the student to learn.)

(iv) Note John interrupting Sarah, who has raised her hand (page 3); Bill doesn't address this. This could raise male TA/female students/equity issues.

(v) There are other problems which could arise in the context of an interactive discussion section, for example classes where no one answers, or classes with a single dominant person. One can launch a discussion by asking how Bill (or how the participants) would react to these problems.

2. *How does Bill balance his concern about "just working the problems" for the students with his desire to be responsive to their needs?*

   a. What do you think is the correct balance?

   b. Do you agree with Bill's TA training course, where "they made a big deal about NOT just doing the problems for the students"?

   c. How can he make doing the problems a true learning experience?

3. *How do you think Bill should apportion his time in the next section? How should he prepare for that section?* What would you do? Note the issue of 'keeping up with the syllabus'.

4. *Should Bill try to make his section useful for the entire class?* To what extent should Bill tailor his section for Sarah, who has done her homework and so is bored (but who might respond to a serious effort to challenge her academically and to stimulate her interest in mathematics), and for Jim, who is struggling with the material?

5. *How should Bill follow up with individual students after this week's section?* What should Bill say to Jim, Fred, Kathy, or Sarah if he saw one of them in the hall after class?

III. Wrap-up Questions

1. *Should Bill be so upset at the conclusion of the case?* Why? Is there anything you could tell him that might make him feel better?

2. *Are there any concrete suggestions you could make to Bill about how he ran the section which might help him in the future?* In answering this question, imagine that you had observed the discussion section described here. Remarks:

   (i) The issue of student expectations for section arises here–what do students expect section to offer them?

   (ii) Also note the need for more TA/prof contact; setting student expectations for section would best be done with in coordination with the prof.

   (iii) Bill needs to encourage questions.

   (iv) The scenario raises the question as to how Bill's students have done on the past homework. Can he find out?

3. *How do you think Bill should spend the rest of the section?* What should he do after it is over? (One suggestion is that Bill should consult with the prof in charge of the course.)

4. *How should Bill prepare for future sections?* It may be useful to emphasize how crucial preparation is.

Exercise for: The Quicksand of Problem Four

Imagine that you are the faculty member in overall charge of Bill's Calculus course. As part of the standard departmental procedure, you must visit his class and complete the form below.

Departmental TA Review Form

Please visit each of your TA's sections during the first two weeks of the semester and complete the following form based on your evaluation of his or her classroom performance. Discuss your evaluation with your TA. Both you and the TA should sign this form. Return it to the Director of Graduate Studies.

This information will be used by the Director of Graduate Studies when making decisions regarding support and assignments of TA's.

For each quality, rate the TA as poor, good, or excellent. Please support your answers with an example if possible!

1. The TA was prepared for class.

2. The TA answered students' questions clearly.

3. The TA's mathematical explanations were clear and correct.

4. The TA used the blackboard or overhead projector effectively.

5. The TA's English was sufficiently clear.

6. Most of the students were present at the section.

7. Give your impression of the overall classroom atmosphere – did the students seem interested, involved, bored, lost?

8. Are there any concrete suggestions for improvement you would like to give the TA?

We have reviewed this report together:

Professor's signature: TA's signature:

Salad Days
Teaching Guide

Case Synopsis

Evan is a well-intentioned, mostly hard-working, student who is close to failing his calculus class. He is stretched too thin among many activities. He seeks relief mid-way through the semester in the form of an extension on a project, is rebuffed by his TA, and continues to do poorly. After seeking help from the TA again late in the semester in a bumbling way, Evan realizes that he must choose between withdrawing from the class, with possible financial aid consequences, or trying to squeak by with a low grade.

Key Issues in this Case

1. When and to what degree it is appropriate to intervene with struggling students.
2. How this is best accomplished.
3. Dealing with students who are behind and want extensions.

Secondary Issues

1. Estimating the time burden a class places on students.
2. Providing timely and accurate feedback about student grades.
3. Delivering unpleasant news in an appropriate and professional way.
4. The consequences of assigning projects whose technologies/means of accomplishment are outside the instructional scope of the class.

Initiating the Discussion

Have participants analyze the two characters in the Case and their behavior. Have them decide if Arnold's behavior is fair, what other options he might have had, and which among them is the best. Have them describe what they would have said to Evan in both of the conversations described in the Case.

The discussion leader may wish to specify the context in which Arnold and Evan are interacting (for example Evan is one of a hundred students who meet with Arnold in recitation sections once a week, or Evan is one of twenty-five students in a class Arnold teaches), to match the institutional context where the Case is being used.

Wrap-Up Suggestions

Have participants practice talking with a student such as Evan about his performance in his math class.

In closing or before, provide the group with written information about resources for struggling students at your institution. It may be useful to also provide them with information about institutional withdrawal, failure, and retake policies and deadlines.

Additional Discussion Questions

1. Is Arnold's response to Evan's request to hand the project in late appropriate? Is there anything you would have done differently in Arnold's position?

2. What might Arnold have done to prevent the final situation? (The group could generate a list in response to this question. For example, Arnold could have looked up Evan's grades the first time he talked to him, and he might then have responded differently, alerting Evan more directly to his academic jeopardy. He might also have intervened after the second exam or after he graded the student projects.) It might be useful here to hand out materials about what resources exist to help struggling students.

3. What are techniques for alerting weak students to their risk of failing the course and getting them to make changes? (For example, some instructors ask the student to make a written plan rather than making a plan for them.)

4. Characterize Evan as a student. What is Evan's probable attitude toward Arnold? How would Evan describe Arnold, and should anyone care?

5. Suppose that Arnold was privy to some of the turmoil in Evan's life, his job and roommate pressures. What, if anything, would change? What, if anything, should change?

6. Suppose Evan was an obviously bright student, who was clearly underperforming by getting B's; would your answer to question 5 be different? If so, how?

7. What should you do if you are concerned about a student's mental health? What would raise such concerns on your part?

8. How typical is Evan of your institution? How many students have large external job or internship requirements? How varied are the backgrounds and majors of students in your classes?

Seeking Points
Teaching Guide

Case Synopsis

Daniel, an advanced graduate student, is teaching a section of Calculus I. After the first examination is handed back a surly student, Sam, comes to Daniel's office. Sam believes that he is unfairly penalized when he does not receive full credit for using the power rule, which has not yet been taught in the class, to answer an examination question which requires a derivative, rather than computing the limit of the difference quotient.

Key Issues in this Case

1. Why teach the difference quotient?
2. How does one communicate these reasons to the students?
3. How does one test understanding of the difference quotient?
4. Fairness and grading.

Secondary Issues

1. Teaching Calculus to students who have already studied it in high school.
2. The roles of practice exams and of review sessions.
3. Dealing with surly students.

Initiating the Discussion

Break into small groups and ask each group to list the issues and conflicts in the Case and to identify the most important ones. Then consolidate on the blackboard or overhead.

Continue by asking questions from the list on the next page, based on the issues identified as most important. To avoid missing a main issue, one could begin by asking participants to articulate reasons why one *should* teach the difference quotient; then ask them how to communicate these reasons to the students.

Wrap-Up Suggestions

Ask each student for one thing which they would do as Calculus teachers to avoid Daniel's situation, or else get students to summarize main aspects of the discussion.

Possible Pitfalls

New TAs may not have thought through the key issues above, and may not in fact be able to articulate good reasons to teach the difference quotient. If you are using this Case with a group of graduate students who do not have much teaching experience, it would be extremely helpful to recruit an experienced TA or an additional faculty member to join in the discussion.

135

Additional Discussion Questions

I. Communication of the Mathematics

1. Why should we teach the difference quotient?
2. How can we explain why we teach the difference quotient to students? (Note that many students have access to computer algebra systems which will differentiate many functions with the click of a mouse.)
3. How can we test students' understanding of the difference quotient and of the definition of the derivative? Write an examination question which does so.
4. How does Sam understand the definition of the derivative? What could you say to him to strengthen his understanding?
5. How would you respond to Sam's comment that the desired answer to Problem 2 is "Just different formulas"?
6. Is there a difference between Sam's gap in understanding and Daniel's thoughts about flat descent and the paper he is studying?

II. Teaching and Classroom Issues

1. Should Daniel give Sam more points? Why or why not?
2. What issues arise in teaching Calculus to students who have already studied some of the material in high school? How can one address them?
3. What is good use of a practice exam? Should one give students questions which are close to the actual examination questions or does this undermine the study process?
4. What is good use of a review session? Should one give information about an examination in the review session?
5. Evaluate Daniel's handling of his conversation with Sam. Could he have probed Sam's understanding? Should he follow up with Sam in any way?
6. What can one say in handing back an examination in which many students will be disappointed with their performances?
7. Should Daniel address Sam's concerns in class the next day? If so, how can Daniel explain the grading policies to his class?
8. How can one write an examination where students and instructors believe that the same answers are correct, one which does not allow for similar misunderstandings?
9. Would you respond to the situation differently if the student complaining had been someone who had attended many office hours and who you knew had worked very hard? Someone who you believed did in fact know the material but missed the review session due to an orchestra rehearsal? Someone whose confidence was shattered by their low score? Someone who was crying?

10. What steps should Daniel take the next time he teaches Calculus I so that he can avoid the situation which he finds himself in here?

Exercise for: Seeking Points

Imagine that you are Daniel, the TA in the case study. You receive the following email.

From: Ruth Allen <rallen@uhp.state.edu>
To: Daniel Baker <dbaker@math.state.edu>
Re: Sam Epstein

Dear Mr. Baker,

I am the coordinator of the University Honors Program. Your student, Sam Epstein, has been nominated for the Meade scholarship, which is given every year to an outstanding freshman student in Mathematics and the Sciences. As you are his Calculus teacher, we on the evaluation committee would find it very helpful if if you would write a brief letter giving your impression of Sam's performance in your class up to this point.

Thank you very much.

Sincerely,

Ruth Allen

What would you write in such a letter?

Study Habits
Teaching Guide

Case Synopsis

Angelica, a graduate student from another country, is teaching a section of second semester Calculus. Six weeks into the term she is enjoying her role as 'professor'. However, there are limitations to her independence as the syllabus has been provided and the hour tests and final exam are common to all sections of the course. Angelica has high expectations for the students in her class and is concerned over their poor performance to date. She decides she would like to make some modifications in the course operation.

Key Issues in this Case

1. Motivating students to study and do homework.
2. The role of examinations and other methods of assessment in effective teaching.
3. Effective classroom management.
4. Professor/graduate instructor interactions.

Secondary Issues

1. Teaching sections of a multi-section course with a set syllabus and common examinations.
2. International graduate student expectations of U.S. undergraduates.
3. Balancing graduate student teaching duties with graduate studies.

Initiating the Discussion

Divide the participants into groups of 2 or 3 and ask each group to prepare a list of issues or concerns and to prioritize their list. Then compile a common list on the chalkboard or overhead. If needed, provoke groups to look at both positive and negative aspects of the characters in the Case. Solicit concrete suggestions for improvements.

Wrap-Up Suggestions

Summarize the group discussion briefly. Then ask participants to identify the one feature of the Case that had the most impact on them, or the one concrete teaching suggestion which arose in the discussion which seems most useful.

Possible Pitfalls

This Case carries more immediate meaning for graduate students who are currently in or soon will be in teaching situations similar to Angelica's. If the group consists primarily of TAs who are responsible for problem/discussion sections then the facilitator may need to help the TAs link the issues in this Case to their own teaching.

Additional Discussion Questions

I. Focus on Angelica's Story

1. Angelica wants her students to learn more. To achieve this, she wants them to work harder, and in particular to come to class better prepared. Do you think her strategy of giving more quizzes would accomplish this? What other strategies could she have employed with these aims in mind? (A wide range of responses is possible here, depending on institution and participants. A few possibilities are: working with the professor and other TAs to set higher standards, increasing the motivation given the mathematics taught in the course and explaining the need for strong knowledge of the material for subsequent work to the students, posting materials on the web, group work, assigning one-minute papers to sample student knowledge and set expectations, publishing student homework solutions, having a mentor visit her class and make suggestions, and asking more questions of her students both in class and in office hours, helping both to decipher their experiences with her course and to ground changes in a realistic framework.)

2. Did Angelica present her proposed changes to the class effectively? Are there other ways she might have tried presenting them? Also, was Angelica out of order in wanting to change grading policy midcourse?

3. Are the student responses to Angelica's speech about initiating quizzes ones you would anticipate? Are they typical and fair? Are there other student voices you would expect to hear?

4. What, if anything, should Angelica do after overhearing a negative comment about her teaching? What is your perception of Angelica as a teacher?

5. Do you agree with Kathy's comments? If you were Kathy, what would you have said to Angelica?

6. Why do you think Professor Jacobs encouraged Angelica to seek students' opinions regarding her plan to initiate more quizzes? Was this a good suggestion? What else could have been suggested?

7. Is Professor Jacobs adequately fulfilling his role as course coordinator? Should he be more sympathetic to Angelica's concerns?

8. Angelica is willing to devote a substantial amount of extra time to her students if it will help them learn, while Professor Jacobs suggests that she not "overdo it". What is an appropriate balance between one's own studies and one's teaching responsibilities?

II. Focus on Teaching

1. What are strategies for getting students to study more, do more homework, and engage more with the material? (See question 1 above.)

2. What is the role of examinations and other methods of assessment in effective teaching?
3. In addition to one hour tests, what are other ways to assess student performance?
4. What advice can one offer teaching assistants so that they can perform their teaching duties effectively and have adequate time for their own graduate studies?

III. Focus on Related Matters

1. What input should graduate students have in the operation of courses in which they are to serve as instructors?
2. What guidance can be offered to graduate students who did not receive their undergraduate education in U.S. schools? What information should be shared about typical U.S. student behavior and expectations?

Studying the Exam
Teaching Guide

Case Synopsis

This case presents sample test questions and asks participants to make a 50 minute midterm. There are 3 versions: College Algebra, Calculus II, and Multivariable Calculus.

Key Issue in this Case

Writing a good examination. This includes probing student understanding and misunderstanding, writing clear questions, fairness, time concerns, grading concerns, and use of technology.

Initiating the Discussion

Have the students carry out the exercise, either in class or before they arrive; in advance is preferable. Poll the students and compare and contrast responses. Refer to the detailed teaching notes for each case for a list of comments on the given questions.

It is valuable to emphasize the question *"what does this question test"* as the group analyzes the given questions or discusses their own.

Wrap-Up Suggestions

Summarize the issues which arise in writing a test; see the instructions to the case itself for a good start.

One could also ask the group to write an exam question which tests understanding of a specific topic (such as the difference quotient, the definition of the definite integral by Riemann sums, or the fundamental theorem of calculus), and compare responses.

As a follow-up activity, one could address the issue of making a grading curve by giving students a set of examination scores and asking them to assign grades. One could also ask students to work with another version of this case.

145

Possible Pitfalls

1. Making an exam is hard work, and graduate students are likely to find it such. It is worth emphasizing that learning the principles now will save them considerable time later. It is also worth noting how important grades are to most undergraduates.

2. Graduate students sometimes forget how much quicker they are than the students they teach, and don't appreciate how difficult some undergraduates find mathematical tricks.

3. Graduate students may not appreciate the importance of giving exams which separate out the C's from the F's as well as the A's from the B's. If participants create exams which are too difficult for most students, the case facilitator may wish to pose a hypothetical distribution of student scores with the majority very low and ask how grades should then be assigned.

4. Graduate students may be most interested in a course similar to the ones they are involved with. In a diverse group some graduate students may find themselves working with a course which is not of immediate relevance.

Studying the Exam, College Algebra Version
Teaching Guide

1. Instructions need to say "Solve exactly", if the approximate answer obtained by tracing using a graphing calculator is not to be given credit.

2. This problem is likely to be difficult to grade. Some students are likely to get

$$x = \frac{1}{2}\left(\frac{\log(0.2)}{\log 5} + 1\right),$$

rather than 0. How many points should this be given? One could say "solve exactly and simplify the answer", but must still be prepared to grade an expression such as that above.

Students may feel that this problem is a "trick", as getting the answer 0 involves seeing that $0.2 = 5^{-1}$ or that $\log(0.2) = -\log 5$. Note also that students with graphing calculators may be able to solve the problem without doing this.

3. This problem will be very hard if students haven't seen quadratics with a^x as the variable. Even if they have, recognizing $4^x = (2^x)^2$ and $2^{x+1} = 2 \cdot 2^x$ will be difficult for many students. This problem can also be done by tracing on a graphing calculator.

4. This problem can be done by linear regression on a calculator, so it either needs to be given on an exam where calculators are not allowed, or rewritten with letters, or one needs to ask for an explanation of the steps. This last option is less satisfactory as some students may write that they used their calculator, and even explain what routine, and this is the answer they got. It is generally better to avoid putting oneself in a position where such arguments are possible.

5. Students may assume that $a \neq 0$.

6. This problem will be hard for some students because the equation of the line is in an unfamiliar form, and because of the phrase "double the x-intercept". Need to assume $b \neq 0$.

7. The first three graphs can be done automatically on a calculator; the fourth cannot. The relationship between the equations suggests that the writer wanted the student to reflect on the relationship between the graphs. However, this is never asked explicitly, and so will probably not happen. If that is wanted, it should be asked for explicitly. Grading explanations is time consuming, but often illuminating (for both teacher and student).

8. Apart from the rather dubious assumption that a linear trend can be based on two years of data, this problem involves setting up a linear function, which students should be able to do. The other aspect which may cause confusion is the use of 80 per 1000, instead of 8 per 100 or 8% . (This was in the original article, and is common in medical literature.)

9. This problem will be hard because of the confusion between whether the lines cross and whether the ship and boat collide, and because the speed is measured along a diagonal line. In addition, the question needs to say that the ships are moving towards one another.

10. True-false questions like this will strike many students as hard at first. However, such questions can be useful in exposing students' misunderstandings, especially if students are asked to explain or justify their answers.

The initial question, "Are the following statements true or false?" should be reworded to avoid the answer "Yes."

Studying the Exam, Calculus II Version
Teaching Guide

1. A very hard problem for most students. The function 2^α and the variable α will both strike students as strange. In addition, the double use of integration by parts is difficult.

2. Standard problem.

3. Not hard, and the set up to make the substitution is obvious.

4. Involves what the students will think of as a "trick": looking at $t^3 e^{t^2}$ as $t^2 \cdot t e^{t^2}$, or making the substitution $u = t^2$. If students have seen this in class, this is a reasonable question. In most institutions, it probably would not be wise to give a question like this for the first time on an exam.

5. Standard problem.

6. Since the graph of the function $y = x^2 - x$ lies below the x-axis for $0 < x < 1$ and above for $x > 1$, this problem is likely to bring up the distinction between the integral and the area. Note that it would make a difference if the function involved was $y = x - x^2$: since $\int_0^2 x^2 - x \, dx = 2/3$, and $\int_0^2 x - x^2 \, dx = -2/3$, changing the function in this way makes it less likely that students will give the same answer to parts (a) and (b).

7. The fact that this function, $y = x^2 - x$ crosses the x-axis at $x = 1$ may cause confusion for better students – someone who plugs into the formula $\pi \int_a^b y^2 dx$ without thinking may do better than someone who thinks about it. This is something to be concerned about!

8. This is a hard problem because there is no formula to be plugged into and the region is not easy to visualize. But it is a good problem to test student understanding.

9. and 10. Leaving out parts (a) and (b) in this problem changes the focus and the level of difficulty in this problem enormously. In #10, students have to pick the strategy as well as do the computation. Many students will find it hard to know how to start #10, although one could argue that picking a strategy is the point of the problem.

11. Part (a) is likely to be hard to grade because it says "show" and the answer is given. However, if part (a) is omitted, students who cannot set the integral up are shut out of attempting part (b).

Students will probably have difficulty evaluating the integral exactly. Although the form of the integral, $\int \frac{A}{e^{kt}} dt$, is not particularly hard, the constants are messy enough to throw many students off track. Many students who get the answer correct will do so by estimating the improper integral numerically.

Studying the Exam, Multivariable Calculus Version
Teaching Guide

1-3. There is nothing that requires the student to understand the meaning of a partial derivative. However, the problems are otherwise quite reasonable.

4. Some students, perhaps many, will not see the relationship between parts (a) and (b). How do you grade part (b) for a student who substitutes $x = 1.1$, $y = 1.98$ into $f(x, y)$? Making the function "messier" will not help if students have calculators. Either way, how do you respond to the argument that "This is a dumb problem, as the best thing to do is substitute"?

5. This problem will be very hard if students have not yet seen local linearity/Taylor approximation. (Same comment applies to Problem 4, although there students can find a way around using local linearity.) If students have not seen these ideas, they may feel the problem is unfair, as it gives a large advantage to students who have taken the course before.

6. Differentiating this function will be hard for many students for two reasons. The use of c as a variable will confuse some. In addition, the function will seem complicated.

7. Inserting "at the point" before $(1, 2)$ will made this question clearer.

8. The phrase "in the direction of $(1, 2)$" is ambiguous; does it mean toward the point $(1, 2)$, or in the direction of the vector with components $(1, 2)$?

9. Students have surprising difficulty keeping straight which quantities are vectors and which are scalars. This question should be reworded to avoid the possibility of students answering "Yes" to the question "Are the following quantities vectors or scalars?"

10. Although this problem is probably intended to check the multivariable chain rule, some students may substitute for x, y, and z in the expression for w and then differentiate using single variable techniques.

11. This question avoids the substitution problem in # 10. However, the lack of formulas for the functions will make this problem seem difficult to some students.

12. Part (a) concerns the interpretation of the gradient; it is a reasonable question for most classes. Part (b) will strike many students as hard.

13. The algebra in problems involving the use of Lagrange multipliers can be a real mine field and can make grading difficult. In this case, various algebraic tricks can be used to simplify the algebra – however, many students will not think of them, and one must be prepared to grade those papers.

There's Something about Ted

Part II

Ted went back to his office. A few minutes later, he emailed his old grad school roommate, who was also an instructor at a big state university, describing the situation. His roommate suggested handing out a survey to the class. Ted thought that this was a good idea. He found one of the standard OU student survey forms lying in his office. It had eight questions, asking the students to rank several factors on a scale of poor, fair, good, very good, excellent. The questions about the instructor asked:

1. Does the instructor have a firm command of the material?
2. Does the instructor communicate the ideas clearly and effectively?
3. Does the instructor give fair quizzes and exams?
4. Does the instructor show concern for his or her students?
5. Rank the instructor overall.

The form had a space at the bottom for the students to fill in general comments.

In class the next day, Ted distributed the surveys to the 18 or so students who had showed up (out of the 30 registered for his course). He collected them, and, as he walked back to his office, flipped through the forms. The numbers looked pretty bad (except on question 1 – the students seemed to have the impression that he knew Calculus!) but what really shocked him were the comments:

```
I am so frustrated.  I am working as hard as I can while holding
down a 20 hour a week job.  I'm sorry that I'm not doing as well
as the HSU students, but I'm really trying hard, and I wish you
would help instead of criticize.
```

```
Go back to HSU, big shot.
```

```
This class is a ripoff.  It takes twice as much work as any
other class I'm taking, but the curve on the midterm was really
unfair.  It's just not fair to work so hard and get such a poor
grade.
```

```
The professor spends the entire class writing formulas on the
board, so fast that I can barely keep up with him.  I don't know
```

what he's doing or why. Please slow down and explain a little more!

We are learning lots of facts and formulas, but I don't see how they fit together.

Professor not like questions. slow him down.

The lectures don't add anything to the book. The Professor just writes down all the equations from the book.

This class is so boring. Formulas, formulas, formulas. Who cares?

I don't understand why learning Calculus is required to graduate. Why in the world are we doing this?

The professor is showing us a great deal of extra material and he lectures very clearly. A very interesting class.

There's Something about Ted
Teaching Guide

This case is designed to be used in two parts. The second part of the case should be distributed only after discussion of the first part; it is not included in the graduate student edition.

Case Synopsis

Ted, a recent Ph.D. from an elite university, confronts difficulties adapting to teaching at a large state school. In Part I, Ted gets pressure but no help from his department chair. In Part II, most of his class responds to his survey with complaints and some with hostility.

Key Issues in this Case

Part I.

1. What to do when a class is not going well, and when students complain.
2. Setting expectations in terms of classroom behavior and in terms of academic performance.
3. How does one make the transition between one institution and another, in particular how can one evaluate what teaching changes are necessary?

Part II.

1. Communicating the big mathematical picture: what should one communicate in lecture and what should one leave out?
2. Workload and grading issues; setting appropriate standards.
3. Avoiding or changing a hostile class-teacher relationship.

Secondary Issues

Part I.

1. Matching course content with class abilities, backgrounds, and interests.
2. Relations with one's Department Chair.
3. Communication with one's students.
4. Finding a teaching mentor.

Part II.

1. Soliciting feedback from students.
2. Finding a teaching mentor.

Initiating the Discussion

Before beginning, it might be useful to mention to graduate students that though this Case concerns a new Ph.D., many of the issues in it pertain to graduate TAs and TFs as well.

The crisis in this case is apparent, which helps initiate the discussion. For each part, one can ask the participants "What's going on here?" and "What

155

would you do if you were Ted?". It may be helpful to make some lists of answers on the blackboard.

An alternative way to use the case is for the facilitator to play the role of Ted and ask the participants to play the role of peer-mentors.

Wrap-Up Suggestions

Ask each participant for a short answer to a question. Possible questions include: "What would you say at the next lecture?" (after the survey in Part II) and "What one piece of advice would you give to Ted if he were your of-ficemate?". Alternatively, if the discussion has already covered these points one could ask each participant for the idea which arose in the discussion which will be most useful in their own teaching.

Possible Pitfalls

The Case reports on a negative situation, but one would like to keep the discussion focussed on positive remedies.

Additional Discussion Questions

Part I.

1. What would you do now if you were Ted?
2. How could Ted get additional information about his classroom perfor-mance? About what the standard practices for teaching Calculus at OU are?
3. What other support could Ted solicit from the department to help him through this crisis? How should he do this?
4. Defend or critique the statement "Students at OU are entitled to just as complete a Calculus course as those at HSU." If defending, how would you explain to your class why such an in-depth course is important to them? If critiquing, what would you say to OU students who express concern over being inadequately prepared for their future courses?

Part II.

1. What would you do now if you were Ted? What should Ted say at the next lecture? What changes should he make for the rest of the course, if any?
2. If you were Ted's office-mate, what are some concrete pieces of advice you would give him?
3. Should Ted find a teaching mentor? How?
4. Analyze the comments by the students. Are there particular students whose concerns one should address individually?
5. Give an example of how you would motivate a formula from Calculus before deriving it, to avoid comments similar to those Ted received.

What Were They Thinking?
Teaching Guide

Case Synopsis

Hugh Brightman, a second-year graduate student, is teaching his own Calculus II class under faculty supervision after a successful year as a TA for a recitation section in a large class taught by a professor. Although not many students have been coming to his office hours, Hugh is confident that they are well prepared for his first hour exam. He is shocked when he discovers that they don't seem to have learned even the most basic techniques and concepts.

Key Issues in this Case

1. How to read students' silences.
2. Writing a good exam.
3. Dealing with a disastrous exam.

Secondary Issues

1. How to raise the mathematical level of a class and make it interesting.
2. Balancing graduate mathematics studies and teaching responsibilities.
3. The difference between the roles of recitation-section instructor and course instructor.
4. Explaining the concept of half-life.

Initiating the Discussion

Break the participants into groups and ask them to analyze the situation, based on the information provided by the Case. (It may be helpful to state explicitly that the information given is somewhat sketchy and may lead different students to quite different answers.) After a few minutes, ask each group to summarize their answers; collect and organize the conclusions on the board. Then ask students what Hugh should do next.

Wrap-Up Suggestions

1. Have participants list ways to determine how the students in one's class are doing. How can one be sure that one knows if they are lost or bored? If they find the class too easy or too hard?

2. List aspects of writing a good exam, and mechanisms to ascertain the impact of an exam before it is given.

3. Have participants describe what they would say when handing back the examinations in the next class, if they were in Hugh's position, and what they would do afterwards.

Possible Pitfalls

Make sure that the discussion does not polarize, either for Hugh or against him. If necessary, point out that sooner or later most of us give an exam like Hugh's, but that it can be avoided.

Additional Discussion Questions

1. How could Hugh have avoided being so shocked at his students' performance on the exam? What signals were there? What led him to misinterpret them?

2. What should Hugh do next? How should he grade the exam? What should Hugh say to the class as he hands it back?

3. How could Hugh have ascertained the level of his class's background? If some of the students had studied the material before while others hadn't, how could Hugh have adjusted for this?

4. How did Hugh attempt to raise the mathematical level of his class and make it interesting? Was this a good idea? What would you do to accomplish this if you were teaching Calculus II at a similar institution?

5. Why is there such a contrast between Hugh's successful performance as a recitation section leader the previous year and his current problems?

6. Why did Paul advise Hugh to drop part of his question on half-life? Was he right? Have participants write an exam question like the one described in the case (including the dropped part), then have them write a possible student answer to someone else's question.

7. Why did students not answer the question on half-lives the easy way? Did none of them understand the concept? If so, and assuming the class has the average range of abilities, how did Hugh fail to get the concept across? If some of the class did understand the concept, why didn't they use their understanding? Have some participants give explanations of half-life, and others critique them.

8. Who bears primary responsibility for the disaster: Hugh, Professor Gatewood, or Hugh's students? Have participants act out (a) an exchange between Hugh and a student complaining about the exam, and (b) an exchange between Professor Gatewood and a parent complaining about Hugh's performance.

9. Would your analysis of the Case change if the class was an honors section at a large state university? A class of engineering students? If so, how?